T0139799

SEMA SIMAI Springer Series

# ICIAM 2019 SEMA SIMAI Springer Series

Volume 8

This sub-series of the SEMA SIMAI Springer Series aims to publish some of the most relevant results presented at the ICIAM 2019 conference held in Valencia in July 2019.

The sub-series is managed by an independent Editorial Board, and will include peer-reviewed content only, including the Invited Speakers volume as well as books resulting from mini-symposia and collateral workshops.

The series is aimed at providing useful reference material to academic and researchers at an international level.

More information about this subseries at http://www.springer.com/series/16499

Maria Rosaria Lancia • Anna Rozanova-Pierrat
Editors

# Fractals in Engineering: Theoretical Aspects and Numerical Approximations

*Editors*
Maria Rosaria Lancia
Dipartimento di Scienze di base ed
Applicate per l'ingegneria
Sapienza Università di Roma
Roma, Italy

Anna Rozanova-Pierrat
Department of Mathematics
CentraleSupélec, Université Paris Saclay
Gif-sur-Yvette, France

ISSN 2199-3041 ISSN 2199-305X (electronic)
SEMA SIMAI Springer Series
ISSN 2662-7183 ISSN 2662-7191 (electronic)
ICIAM 2019 SEMA SIMAI Springer Series
ISBN 978-3-030-61805-6 ISBN 978-3-030-61803-2 (eBook)
https://doi.org/10.1007/978-3-030-61803-2



This Springer imprint is published by the registered company Springer Nature Switzerland AG.
The registered company address is: Gewerbestrasse 11, 6330 Cham, Switzerland

# Preface

Many natural and industrial processes lead to the formation of rough surfaces and interfaces. Computer simulations, analytical theories and experiments led to significant advances, in modeling these phenomena across wild media. Fractals provide a good tool to describe such wild geometries, as well as in the case of those phenomena that take place in small volumes and large surfaces.

This special issue is devoted to collect new results on fractal applications in engineering from both a theoretical and a numerical point of view, having in mind further developments of fractal and prefractal geometries in industrial applications. Some of these results have been presented during the special Minisymposium MS256 entitled *Fractal Applications in Engineering: Theoretical Aspects and Numerical Approximations*, held in Valencia, Spain, July 15–19, 2019.

We thank all the authors for participating to this special issue as well as all the referees who contributed with their constructive and rigorous reviews. Special thanks go to the Editors-in-Chief Pablo Pedregal Tercero of University of Castilla-La Mancha and Luca Formaggia of Politecnico di Milano for inviting us to edit a Special Issue of the SEMA SIMAI Springer Series.

Guest Editors

Roma, Italy — Maria Rosaria Lancia

Gif-sur-Yvette, France — Anna Rozanova-Pierrat

# Contents

# Editors and Contributors

## About the Editors

**Maria Rosaria Lancia** is Professor of Mathematical Analysis at the Sapienza University of Rome, where she received her PhD in Applied and Theoretical Mechanics. Her current research interests are fractal analysis and numerical approximation of BVPs in fractal domains. The emphasis is on linear, quasilinear and fractional BVPs in and within fractal domains possibly with dynamical boundary conditions and vector analysis on fractafolds. She is an editorial board member of Fractal and Fractional, MDPI, and she has been a member of the Journal of Applied Mathematics and Computation, Hill Publishing Group up to 2019.

**Anna Rozanova-Pierrat** is Associate Professor of Applied Mathematics at the CentraleSupélec, University Paris-Saclay, France. She obtained his PhD on Applied Mathematics at the University Pierre et Marie Currie Paris 6 and RUDN (Moscow, Russia), where she finished her studies on Theoretical and Applied Mathematics. Her current research interests are motivated by physical and engineer problems (such as models of non-linear acoustics, de Gennes hypothesis on the speed of the heat propagation between two media, shape optimization) involving irregular and fractal boundaries.

## Contributors

**C. Alberini** Dipartimento SBAI, Sapienza Università di Roma, Rome, Italy

**Massimo Cefalo** Dipartimento di Ingegneria Informatica, Automatica e Gestionale, Sapienza Università di Roma, Roma, Italy

**Simone Creo** Dipartimento di Scienze di Base e Applicate per l'Ingegneria, Sapienza Università di Roma, Roma, Italy

**Salvatore Fragapane** Dipartimento di Scienze di Base e Applicate per l'Ingegneria, "Sapienza" Università di Roma, Roma, Italy

**Malcolm Gabbard** Colorado College, Department of Mathematics and Computer Science, Colorado Springs, CO, USA

**Mirko Gallo** Dipartimento di Ingegneria Meccanica e Aerospaziale, Sapienza Università di Roma, Roma, Italy

**Maria Rosaria Lancia** Dipartimento di Scienze di Base e Applicate per l'Ingegneria, Sapienza Università di Roma, Roma, Italy

**Carlos Lima** California State University Long Beach, Department of Physics and Astronomy, Long Beach, CA, USA

**M. V. Marchi** Dip. di Matematica, Sapienza Università di Roma, Roma, Italy

**Gamal Mograby** University of Connecticut, Department of Mathematics, Storrs, CT, USA

**Umberto Mosco** Department of Mathematics, Worcester Polytechnic Institute, Worcester, MA, USA

**Luke Rogers** University of Connecticut, Department of Mathematics, Storrs, CT, USA

**Anna Rozanova-Pierrat** CentraleSupélec, Université Paris-Saclay, Saint-Aubin, France

**Alexander Teplyaev** University of Connecticut, Department of Mathematics, Storrs, CT, USA

**Paola Vernole** Dipartimento di Scienze di Base e Applicate per l'Ingegneria, Sapienza Università di Roma, Roma, Italy

**S. F. Vita** Dipartimento di Matematica, Sapienza Università di Roma, Rome, Italy

**Maria Agostina Vivaldi** Dip. Di Scienze di base ed Applicate per l'ingegneria, Sapienza Università di Roma, Roma, Italy

# A Numerical Approach to a Nonlinear Diffusion Model for Self-Organized Criticality Phenomena

**C. Alberini and S. Finzi Vita**

**Abstract** We describe a numerical implementation of a differential model for the simulation of self-organized criticality (SOC) phenomena arising from recent papers by Barbu (Annu Rev Control 34:52–61, 2010; Math Methods Appl Sci 36:1726–1733, 2013). In that singular nonlinear diffusion problem an initial supercritical state evolves in a finite time towards a given critical solution, progressively from the boundary towards the internal regions. The key elements are the Heaviside function which plays the role of a switch for the dynamics, and the initial boundary contact with the critical state.

A finite difference implicit scheme on a fixed grid is proposed for a regularized version of the problem, with the Heaviside replaced by a $C^1$ function, showing the same behavior of the solution: convergence in finite time toward the critical state on every single node, up to any prescribed accuracy, remaining supercritical during all the process.

The use of synchronized spatial-temporal grids with progressive refinements (in the spirit of Mosco (SIAM J Math Anal 50(3):2409–2440, 2018)) simulates the appearance of short-range interactions of an increasing number of particles, speeding up the convergence to the critical solution, and allowing a strong reduction of computational cost.

The results of some numerical simulations are discussed, in one and two dimensions.

**Keywords** SOC phenomena · Nonlinear diffusion problems · Heaviside function · Finite difference schemes

---

C. Alberini (✉)
Dipartimento SBAI, Sapienza Università di Roma, Rome, Italy
e-mail: carlo.alberini@sbai.uniroma1.it

S. Finzi Vita
Dipartimento di Matematica, Sapienza Università di Roma, Rome, Italy



M. R. Lancia, A. Rozanova-Pierrat (eds.), *Fractals in Engineering: Theoretical Aspects and Numerical Approximations*, SEMA SIMAI Springer Series 8,
https://doi.org/10.1007/978-3-030-61803-2_1

# 1 Introduction

This article is concerned with the numerical study of the following evolutive singular nonlinear model:

$$\begin{cases} u_t - \Delta H(u(t) - u^c) \ni 0 & \text{in } \Omega \times (0, \infty) \\ u(0) = u^0 > u^c & \text{in } \Omega \\ 0 \in H(u(t) - u^c) & \text{on } \partial\Omega \times (0, \infty), \end{cases} \tag{1}$$

where $\Omega$ is a bounded subset of $\mathbb{R}^2$, $u^c \in C^0(\overline{\Omega})$ a given target function (the *critical state*), and $u^0$ a supercritical initial datum, while $H$ is the multivalued Heaviside function:

$$H(r) = \begin{cases} 1 & \text{if } r > 0 \\ [0, 1] & \text{for } r = 0 \\ 0 & \text{if } r < 0 \end{cases} . \tag{2}$$

This model was discussed in recent papers by Barbu [4, 5], in relation to the study of the so-called SOC (*Self-Organized Criticality*) processes. The fundamentals of the SOC theory date back to the late 1980s, when Bak, Tang, and Wiesenfeld [1, 2] introduced the sandpile cellular automata model in order to analyze the time behavior of avalanches on a $N \times N$ plane lattice. This model became the prototype for the study of many complex system dynamics where the current state evolves spontaneously towards a critical state. In the model when the local height $h_{ij}$ of the sandpile at the $ij$-site reaches a prescribed critical threshold $h^c$ it becomes unstable, yielding the toppling of grains on the four adjacent sites. This automata dynamics was formalized by Dahr [7], who introduced the toppling matrix $D$: after a toppling in the $k\ell$-site, the height $h_{ij}$ of the sandpile at any site changes according to

$$h_{ij}^{t+1} = h_{ij}^t - D_{ij,k\ell} ,$$

where

$$D_{ij,k\ell} = \begin{cases} 4 & \text{if } ij = k\ell \\ -1 & \text{if } ij \text{ and } k\ell \text{ adjacent sites} \\ 0 & \text{otherwise} \end{cases} . \tag{3}$$

By this rule subsequent topplings can occur, generating avalanches which end as soon as stability is reached again over all the lattice sites ($h_{ij} < h^c$ at any site). In compact form the toppling law can be interpreted as an implicit in time nonlinear finite difference system:

$$h^{t+1} - h^t = DH(h^{t+1} - h^c) , \tag{4}$$

for the vector of the heights $h$ and the matrix $D$, and $H$ is now the standard Heaviside function (not multivalued). Note that $D$ recalls the well-known tridiagonal block matrix coming from the 5-point finite difference Laplacian approximation. It is henceforth not surprising that Carlson and Swindle [6] were able to characterize the continuum limit of the cellular automaton proposed by Bak, Tang, and Wiesenfeld as the solution $u(t)$ of the following singular diffusion equation:

$$u_t = \Delta H(u(t) - u^c), \tag{5}$$

with $u^c$ a given critical state which plays the role of the threshold. Then the evolution of the system toward the equilibrium can be described by two distinct time scales, a low one far from the critical state and a fast one in its neighborhood (corresponding to the avalanche process).

The first equation of (1) is the multivalued version of (5). The subsequent relations in (1) simply state that the initial datum $u^0$ is strictly supercritical inside $\Omega$, and that $u^0$ and $u^c$ assume the same value on $\partial\Omega$ (not necessarily equal to 0). Due to the discontinuity of the Heaviside function, the ordinary existence results are not applicable to an equation as (5). That is why the multivalued setting was necessary to prove in [5] that the solution $u(x, t)$ of (1) exists and evolves spontaneously in time towards the critical state $u^c$ from above; in other words the supercritical region is absorbed into the critical region in a finite time. In [5] such solution is proven to belong to $C([0, T]; L^1(\Omega) \cap H^{-1}(\Omega)) \cap W^{1,2}([0, T]; H^{-1}(\Omega))$ if the initial datum $u^0 \in L^2(\Omega)$; it is in $L^\infty((0, T) \times \Omega)$ if $u^0 \in L^\infty(\Omega)$.

It is interesting to note that the three typical aspects of the original sandpile model, i.e. local equilibrium, threshold activation, and diffusive character of the updating rules which cause the solution of (1) in the above settings to evolve in time towards the critical state, can be shared by simple biological models of contamination and epidemic spreading, where the Heaviside term acts as a switch of the process itself (see e.g. [8]). As we will see also in our numerical simulations, the temporal evolution of the solution $u$ is characterized by a progressive alignment to the target function $u^c$ starting from the boundary of $\Omega$, and proceeding towards the more internal values.

In order to approximate problem (1) by a finite difference implicit approach similar to (4), it was first of all necessary to replace the multivalued Heaviside function with a smooth approximation of it: this will be discussed in Sect. 2, together with some properties of this regularized problem.

The numerical scheme was implemented in Sect. 3 on a fixed space grid. Since the discrete settings cannot reproduce completely the fine spatial interactions of the continuous model, the finite time convergence corresponds to the time evolution of the numerical solution to the target on every single node up to any prescribed accuracy, globally with exponential rate (as numerical tests suggest, see Sect. 5, Table 3a), remaining supercritical during all the process.

In Sect. 4 we will adapt our scheme to a space-time synchronized family of grids in order to reduce the convergence time and the total computational cost. The idea was inspired by a recent paper of Mosco [9], where a fully discrete analytic model allowing for infinite degrees of freedom was studied to preserve the physical aspect of SOC phenomena which are intrinsically discrete particle processes, better captured by discrete equations on an infinite spatial-temporal lattice incorporating arbitrary short-range and high frequency particle interactions. The equations of the process on each finite spatial grid are coupled in time with an impulsive change of the initial data at each refinement, keeping constant the parabolic ratio between the discretization steps, up to the desired accuracy.

In Sect. 5 we finally present the results of some numerical simulations both in one and two dimensions.

## 2 The Basic Model

In view of the discretization of problem (1), we first of all introduce a regular approximation $\eta \in C^1(\mathbb{R})$ of $H$ in a neighborhood of the origin, for example by means of the cubic polynomial:

$$\eta_1(r) = \begin{cases} 1 & \text{if } r > \frac{1}{n} \\ -\dfrac{n^3r^3}{4} + \dfrac{3}{4}nr + \dfrac{1}{2} & \text{if } -\dfrac{1}{n} \leq r \leq \dfrac{1}{n} \\ 0 & \text{if } r < -\frac{1}{n} \end{cases} . \tag{6}$$

With respect to the integer parameter $n \in \mathbb{N}$, we see that $\eta_1(r) = H(r)$ for $|r| \geq \dfrac{1}{n}$, and that

$$\|H - \eta_1\|_{L^1(-1,1)} = \int_{-1}^{1} |H(r) - \eta_1(r)|\, dr = \int_{-1/n}^{1/n} |H(r) - \eta_1(r)|\, dr \to 0 \quad as\ n \to \infty .$$

Note that $\eta_1(0) = 0.5$, so that this will be the value taken at the contact points between the solution of problem (1) (with $H$ replaced by $\eta_1$) and the critical one, in particular at the boundary. Moreover, the values of $\eta_1$ in $(0.5, 1]$ characterize the supercritical states of the solution, those in $[0, 0.5)$ the subcritical ones.

There are of course other possible candidates for the approximation of $H$. For example the following monotone functions

$$\eta_2(r) = \frac{1}{(1 + e^{-nr})}, \text{ or } \quad \eta_3(r) = \frac{1}{2} + \frac{1}{\pi} \arctan(nr) \tag{7}$$

have similar properties for a large $n$, even if in those cases the support of $(H - \eta)$ is unbounded. The numerical tests show that the choice of $\eta$ slightly influences the qualitative behavior of the solution, that is the way it converges to the target (more impulsive for $\eta_1$, more smooth for the other choices), but it is not relevant for the final time of convergence, which is essentially determined by the quantity $\eta'(0)$ (in the previous examples: $\eta_1'(0) = 3n/4$, $\eta_2'(0) = n/4$, $\eta_3'(0) = n/\pi$): the highest is this value, the closest we are to the real Heaviside, and the fastest is the convergence to the target (see Sect. 5).

The problem to solve then becomes:

$$\begin{cases} u_t - \Delta\eta(u - u^c) = 0 & \text{in } \Omega \times (0, \infty) \\ u(0) = u^0 > u^c & \text{in } \Omega \\ u = u^c & \text{on } \partial\Omega \times [0, \infty), \end{cases} \tag{8}$$

where $\eta$ denotes one of the approximate Heaviside functions previously defined, for a fixed large parameter $n$ that for simplicity from now on we neglect.

If we set $w\,(x, t) := u\,(x, t) - u^c\,(x)$, problem (8) can also be written as

$$\begin{cases} w_t - \Delta\eta(w) = 0 & \text{in } \Omega \times (0, \infty) \\ w(0) > 0 & \text{in } \Omega \\ w = 0 & \text{on } \partial\Omega \times [0, \infty) \end{cases} . \tag{9}$$

Applying the divergence theorem, together with the boundary conditions on $w$, we get

$$\int_\Omega \Delta\eta\,(w)\; dx = \oint_{\partial\Omega} \nabla\eta\,(w) \cdot \vec{n}\; ds = \oint_{\partial\Omega} \eta'(w)\nabla w \cdot \vec{n}\; ds = \eta'(0) \oint_{\partial\Omega} \frac{\partial w}{\partial n}\; ds \le 0, \tag{10}$$

since $\eta'(0) > 0$ ($\eta$ is increasing at the origin) and the normal derivative of $w$ on the boundary in the direction of the external normal vector is necessarily negative (or eventually zero), since $w \ge 0$ and it decreases to zero at $\partial\Omega$.

From (9) and (10) then follows:

$$\int_\Omega w_t\; dx = \int_\Omega \Delta\eta\,(w)\; dx \le 0\,; \tag{11}$$

according to that, if we define the two quantities:

$$M(t) = \int_\Omega w(x, t)\; dx, \quad E(t) = -\int_\Omega \Delta\eta(w)\; dx\,, \tag{12}$$

then we have $M'(t) \le 0$ and $E(t) \ge 0$. In other words $M(t)$, which can be interpreted as the total mass of the problem (that is the global distance from the target), always decreases with time. This fact does not imply that $w$ decreases at any point of $\Omega$: from (9) we see that this should be equivalent to say that $\eta(w)$ be globally superharmonic in $\Omega$. Of course, according to the specific given initial data, this could not be the starting situation. In fact the second quantity $E(t)$, which represents a sort of distance from harmonicity for $\eta(w)$, might initially grow before it starts to decrease toward zero. We will find the same behavior even in the discrete settings.

For the original continuous model (1) it was proved in [5] that $u(t) \to u^c$ (and $M(t) \to 0$) in finite time. Asymptotically, model (8) shares the same properties. In particular at any point of $\Omega$ as $t$ grows: $w(t) \to 0$, $\eta(w(t)) \to 0.5$, and $\Delta\eta(w(t)) \to 0$. We will prove these properties for the finite difference solution of an implicit scheme which discretizes (8).

## 3 Numerical Approximation on a Fixed Grid

Let us assume for simplicity that $\Omega = [-1, 1]^2$, that is the unit square in $\mathbb{R}^2$. On $\Omega$ we define for $m \in \mathbb{N}$ a sequence of uniform discrete grids $G_m$ of increasing cardinality and size

$$h_m = 2^{1-m}.$$

Then $G_m$ will have $M_m = (2^m - 1) \times (2^m - 1)$ internal nodes over a total number of $(2^m + 1) \times (2^m + 1)$. Note in particular that $G_m \subset G_{m+1}$, so that if a node is in $G_m$ it will belong to every $G_s$ with $s > m$. Concerning time discretization, on the grid $G_m$ we adopted the time step

$$dt_m = 4^{-m},$$

in order to keep constant the parabolic ratio $\gamma$ between the steps, for any $m$: $\gamma = \frac{dt_m}{h_m^2} = 1/4$. Different step ratios can be adopted (up to $\gamma = 0.5$); with larger values, despite the implicit scheme adopted, some instability phenomena occur, due to the nonlinearity and singularity of the problem: the solution can overcome the target at some node, causing loop oscillations and the loss of convergence.

We call $x_{ij}$ the generic node belonging to the $i$-th row and the $j$-th column of the square mesh $G_m$, and $t_m^k = k dt_m$ the instant times where we compute the solution: then by $(u_m^k)_{ij}$ (or simply by $u_{ij}^k$ when $m$ is fixed over all the computation) we denote the discrete solution at time $t_m^k$ in a node $x_{ij}$ of $G_m$. The initial data will be given by

$$u_{ij}^0 = u^0(x_{ij}),\ u_{ij}^c = u^c(x_{ij}), \text{ with } u_{ij}^0 > u_{ij}^c, \text{ and } u_{ij}^0 = u_{ij}^c \text{ on boundary nodes.} \tag{13}$$

A possible numerical solution of (8) on the grid $G_m$ for a fixed $m$ is the solution of the following nonlinear implicit finite difference scheme:

**Scheme (S1)** *Until* $\|u^k - u^c\|_\infty < tol$, *for any* $k = 0, 1, \ldots,$ *for any* $x_{ij} \in G_m$, *solve* :

$$\begin{cases} u_{ij}^{k+1} = u_{ij}^k + dt_m \Delta_m \left(\eta_{ij}^{k+1}\right) := u_{ij}^k + \gamma \left(\eta_{i+1,j}^{k+1} + \eta_{i-1,j}^{k+1} - 4\eta_{i,j}^{k+1} + \eta_{i,j+1}^{k+1} + \eta_{i,j-1}^{k+1}\right) \\ u_{ij}^{k+1} = u_{ij}^k = u_{ij}^c \quad \text{at } x_{ij} \in \partial G_m, \end{cases} \tag{14}$$

where $tol$ is a given tolerance parameter, $\Delta_m$ denotes the usual 5-point discrete Laplacian operator over $G_m$, and of course $\eta_{ij}^k := \eta\left(u_{ij}^k - u_{ij}^c\right)$. Note in particular that $\eta_{ij}^k = 0.5$ on $\partial G_m$.

For any time step the implicit system (14) is solved by the vectorial Newton method, finding (more precisely approximating) a vector $z^* \in \mathbb{R}^{M_n}$, such that $F(z^*) = 0$, where $F : \mathbb{R}^{M_n} \to \mathbb{R}^{M_n}$ is defined by

$$F(z) = z - U^k + \gamma A\eta(z - U^c) - \gamma g\,, \tag{15}$$

and $U^k$, $U^c$ denote respectively the vectors of the values of $u^k$ and $u^c$ at the internal nodes of $G_m$ (taken in the lexicographic order), $A$ is the block tridiagonal matrix associated to the 5-point discrete Laplacian on that mesh, and $g$ is a correction vector which takes into account the boundary values of $\eta$ where $u = u^c$. In practice for any internal node close to the square sides we have to set $g = 0.5$, $g = 1$ near the four vertices, and $g = 0$ elsewhere. Since we have chosen a regular function $\eta$ we may assume $F \in C^1$, so that its Jacobian matrix $J_F(z)$ is well defined, and the Newton method becomes

$given\ z_0 = U^k,\quad for\ n = 0, 1, \ldots, itmax:$

1. $solve\ the\ linear\ system:\ J_F(z_n)\rho_n = -F(z_n).$
2. $set: z_{n+1} = z_n + \rho_n.$
3. if $||z_{n+1} - z_n|| < \tau\ STOP\ and\ set:\ U^{k+1} = z_{n+1}.$

Here $||.||$ denotes the euclidean norm in $\mathbb{R}^{M_n}$, $\tau$ is an assigned tolerance, and $itmax$ is the maximum number of iterations.

We want to prove that the discrete solution of our scheme satisfies the same evolution properties of the solution of the continuous problem, in particular that it always remains over the target and asymptotically reaches it. For simplicity we will prove such a result in the one dimensional case. What we say can be easily extended to the general 2D case.

So let $\Omega = (-1, 1)$, and $G_m$ the mesh over $\Omega$ with $2^m + 1$ equally spaced nodes $x_j$, with $x_0 = -1$ and $x_{2^m} = 1$. Then the 1D version of (14) reduces to

$$u_j^{k+1} = u_j^k + \gamma\ \delta\eta_j^{k+1} := u_j^k + \gamma(\eta_{j-1}^{k+1} - 2\eta_j^{k+1} + \eta_{j+1}^{k+1}) \tag{16}$$

for any internal node $x_j$, $j = 1, 2, \ldots, 2^m - 1$. For a vector $z = \{z_j\}$ defined over $G_m$ we have set $\delta z_j := z_{j-1} - 2z_j + z_{j+1}$, so that the discrete Laplacian operator applied to $z_j$ has now the form of the second order central difference $\frac{\delta z_j}{h_m^2}$. Of course $\eta_j^k := \eta(u_j^k - u_j^c)$, for the chosen relaxed Heaviside function $\eta$. Note that if we set $w_j^k := u_j^k - u_j^c$, by subtracting $u_j^c$ from both sides of (16) such equation can equivalently be written as

$$w_j^{k+1} = w_j^k + \gamma\ \delta\eta_j^{k+1}. \tag{17}$$

The following result holds true:

**Theorem 3.1** *Let $m$ be fixed; then the solution $u^k = \{u^k_j\}_j$ of (16) on $G_m$ satisfies the following properties:*

1. *for any $k$ and $j$: $u^k_j \geq u^c_j$, $1 \geq \eta^k_j \geq 0.5$ (the solution remains always supercritical);*
2. *for any $k$: $\sum_j w^{k+1}_j \leq \sum_j w^k_j$ (the global distance from the target decreases);*
3. *as $k \to \infty$, for any $j$: $w^k_j \to 0$, $\eta^k_j \to 0.5$ (the solution converges to the target).*

***Proof*** Taking into account the boundary values for $\eta$, any iteration of scheme (16) corresponds to solve the following nonlinear system:

$$u^{k+1} = u^k - \gamma A \eta^{k+1} + \gamma g, \tag{18}$$

where the already defined correction vector $g$ has in 1D all null components except the first and the last (equal to 0.5), and $A$ reduces to the tridiagonal symmetric positive definite matrix with all 2 on the main diagonal and -1 on the sub and super diagonals. Note that $Af = g$, with $f$ the constant vector with all the values equal to 0.5. Then (18) becomes

$$u^{k+1} = u^k - \gamma A(\eta^{k+1} - f), \tag{19}$$

showing that the evolution of the system could stop only when $\eta^k$ tends to $f$ (that is when $u^k$ tends to $u^c$ in all the nodes of the chosen mesh). Moreover, since $A$ is a known monotone matrix, in the worst case, that is when $u^{k+1} \leq u^k$, then, from (19), $A(\eta^{k+1} - f) \geq 0$, yielding $\eta^{k+1} \geq f$: that is, the solution remains supercritical ($u^{k+1} \geq u^c$ at any node) since all the values of $\eta^{k+1}$ belong to the interval [0.5, 1]. Of course, when some component of $u^k$ grows, the corresponding value of $\eta^k$ cannot decrease, for the monotonicity of $\eta$. This ends the proof of point 1.

Adding up (17) for any index $j$ we easily get

$$\sum_{j=1}^{2^m-1} (w^{k+1}_j - w^k_j) = \gamma \sum_{j=1}^{2^m-1} (\eta^k_{j-1} - 2\eta^k_j + \eta^k_{j+1}) = \gamma(1 - \eta^k_1 - \eta^k_{2^m-1}) \leq 0, \tag{20}$$

since $\eta^k_1, \eta^k_{2^m-1} \geq 0.5$. This immediately proves point 2. It also says that if the distance $w^k$ from the target grows at some node it has to decrease at least at another one in such a way that the global distance cannot grow. Note that (20) can be interpreted as the 1D discrete version of (10): on the uniform grid $G_m$, it is enough to approximate the integrals by the repeated rectangle quadrature formula of step $h_m$. The quantity

$$M^k_m = h_m \sum_j w^k_j = h_m \sum_j (u^k_j - u^c_j) \tag{21}$$

then represents the discrete version of the mass $M(t)$ at the iteration $k$ on the grid $G_m$. From points 1 and 2 we see that $M^k \to \overline{M} \geq 0$. To conclude the proof of the theorem it is then sufficient to prove that $\overline{M} = 0$. Such result would easily follow if we could prove that $u^k \to \tilde{u}$ for some vector $\tilde{u}$, because in that case $\eta^k = \eta(u^k) \to \eta(\tilde{u}) = f$, yielding $\tilde{u}_j = u^c_j$ for any $j$. But we only know that the sum of the distances converge. In principle we could have that single components of $u^k$ do not have limit but tend to some loops which do not alter the sum. Such situation can be excluded as a consequence of (20): passing to the limit we see in fact that necessarily both $\eta^k_1$ and $\eta^k_{2m-1}$ have to converge to 0.5. This implies for example that the solution cannot oscillate in $x_2$ (otherwise it should do that also in $x_1$); then even $\eta^k_2 \to 0.5$, and so on, with the same reasoning for any subsequent node. □

In the numerical simulations it is easy to see that the convergence of $u^k$ to $u^c$ is not in general monotone; but we can state the following proposition:

**Proposition 3.1** *Assume that: (*) there exists an index $k_0$ such that $u^{k_0}_j \geq u^{k_0+1}_j$ holds true for any $j$; then for any $k > k_0$ the solution of (16) satisfies:*

1. $u^{k_0}_j \geq u^k_j \geq u^{k+1}_j \geq u^c_j, \quad 1 \geq \eta^{k_0}_j \geq \eta^k_j \geq \eta^{k+1}_j \geq 0.5, \quad \forall x_j \in G_m$ ;
2. $-\sum_j \delta\eta^k_j \geq -\sum_j \delta\eta^{k+1}_j \,, \quad -\sum_j \delta\eta^k_j \to 0$ .

In other words after the iteration $k_0$ all the quantities $u^k_j$, $\eta^k_j$ and even the discrete version of $E(t)$, that is $E^k_m = -\sum_j \delta\eta^k_j / h_m$, converge in a monotone way.

***Proof*** It is clear that if $u^k$ decreases also $w^k$ decreases, and $\eta^k$ does the same, for the monotonicity of the $\eta$ function. So, in order to prove 1, we use an induction argument. Let us suppose that for an index $k > k_0$ the property $u^k_j \geq u^{k+1}_j$ holds true for any $j$; we want to prove (reasoning by contradiction) that also $u^{k+1}_j \geq u^{k+2}_j$ for any $j$. Such result together with (*) will prove 1.

If it is false, it should exist at least an index $j$ such that $u^{k+2}_j > u^{k+1}_j$; then $\eta^{k+2}_j \geq \eta^{k+1}_j$, and (from (16))

$$\delta\eta^{k+2}_j > 0 \,;$$

let us show that in such a case the solution has to grow also at an adjacent node; by the induction assumption:

$$\delta\eta^{k+1}_j \leq 0 \ \ (\text{ since } u^{k+1}_j \leq u^k_j) \,;$$

then

$$\eta^{k+1}_{j-1} + \eta^{k+1}_{j+1} \leq 2\eta^{k+1}_j \leq 2\eta^{k+2}_j < \eta^{k+2}_{j-1} + \eta^{k+2}_{j+1} \,,$$

and $\eta^{k+1}$ has to grow, at least at one of the two adjacent nodes; if it grows at both, let us consider anyway the one where the growth is larger, and assume for example it is $x_{j+1}$. The previous argument shows also that, if $\eta^{k+1}$ grows of a positive quantity $\alpha_j$ at $x_j$, then it has to grow at $x_{j+1}$ of a quantity $\alpha_{j+1} > \alpha_j$; in fact:

$$\delta\eta_j^{k+2} = -2(\eta_j^{k+1} + \alpha_j) + (\eta_{j-1}^{k+1} + \alpha_{j-1}) + (\eta_{j+1}^{k+1} + \alpha_{j+1}) =$$

$$= \delta\eta_j^{k+1} - 2\alpha_j + \alpha_{j-1} + \alpha_{j+1} > 0,$$

so that

$$\alpha_{j-1} + \alpha_{j+1} > 2\alpha_j - \delta\eta_j^{k+1} \geq 2\alpha_j \quad \text{(for the induction hypothesis)},$$

and we can deduce that $\alpha_{j+1} > \alpha_j$ (since $\alpha_{j+1} > \alpha_{j-1}$); but a similar argument can be repeated now at the node $x_{j+1}$, showing that necessarily $\eta^{k+1}$ has to grow also at node $x_{j+2}$ by a quantity $\alpha_{j+2} > \alpha_{j+1}$. In fact:

$$\delta\eta_{j+1}^{k+2} = -2(\eta_{j+1}^{k+1} + \alpha_{j+1}) + (\eta_j^{k+1} + \alpha_j) + (\eta_{j+2}^{k+1} + \alpha_{j+2}) =$$

$$= \delta\eta_{j+1}^{k+1} - 2\alpha_{j+1} + \alpha_j + \alpha_{j+2} > 0,$$

so that

$$\alpha_{j+2} > 2\alpha_{j+1} - \alpha_j - \delta\eta_{j+1}^{k+1} > \alpha_{j+1} + (\alpha_{j+1} - \alpha_j);$$

in such a way we should have that $u^{k+2}$, and then $\eta^{k+2}$ has to grow at all the subsequent nodes up to the node $x_{M-1}$ close to the right boundary (here $M = 2^m$), where

$$\delta\eta_{M-1}^{k+2} = -2(\eta_{M-1}^{k+1} + \alpha_{M-1}) + \frac{1}{2} + (\eta_{M-2}^{k+1} + \alpha_{M-2}) > 0\,,$$

yielding $(2\alpha_{M-1} - \alpha_{M-2}) < 0$, which is impossible for what we have stated before. So our initial assumption has led us to a contradiction, and we can conclude by induction that $u^{k+1} \geq u^{k+2}$, as we wanted to prove.

As $\eta_j^k$ decreases and tends to 0.5 for all $j$, point 2 is an easy consequence of (20) and 1, since $-\sum_j \delta\eta_j^k = \eta_1^k + \eta_{2^m-1}^k - 1$. □

Note that (*) holds true for $k_0 = 0$ if for example $\eta = \eta_1$ from (6) and $\eta_j^0 = 1$ for any internal node $x_j$: in that case at the first iteration the solution cannot grow. If it would happen at a node $x_j$, then necessarily $\eta_j^1 = \eta_j^0 = 1$, but from (16)

$$\delta\eta_j^1 = \eta_{j-1}^1 - 2\eta_j^1 + \eta_{j+1}^1 > 0 \quad \Rightarrow \quad \eta_{j-1}^1 + \eta_{j+1}^1 > 2,$$

which is impossible since $\eta_j \leq 1$ for any $j$. Then $u_j^0 \geq u_j^1 \geq u_j^c$ for any $j$. In (8) the Laplacian of the Heaviside plays the role of a sort of switch for the dynamics, so it is not surprising that nothing changes in the discrete model at a node where $\delta\eta_j = 0$: this happens of course when $\eta$ is constant on the three-point stencil (that is when $\eta_{j-1} = \eta_j = \eta_{j+1}$), or more generally when it is harmonic (linear, in 1D) on it. If $\eta = \eta_1$ and the starting values $\eta_j^0$ on the internal nodes of $G_m$ are all equal to 1, the only values of the solution which initially decrease towards the target (by $\gamma/2$ at each iteration) are the ones at the two nodes close to the boundary: for example at $x_1$ one has $\delta\eta_1 = 0.5 - 2 + 1 = -0.5 < 0$. At the more internal nodes the solution does not change. When the distance to the target becomes lower than $1/n$ the value $\eta_1$ starts to relax monotonically from 1 until it reaches 0.75 (since there $\delta\eta_1 = 0.5 - 1.5 + 1 = 0$). It is at that time that the dynamics is impulsively activated at $x_2$ and the value $u_2^k$ starts to decrease towards $u_2^c$ (where $\delta\eta_2 < 0$), while $(\eta_1, \eta_2)$ relax from (0.75, 1) towards the values (0.667, 0.833) (so that again $\delta\eta_1 = \delta\eta_2 = 0$ and $\delta\eta_3 < 0$). Then $u_3^k$ starts to move, and so on. One node at the time, from the boundary towards the interior of $\Omega$, all the values of the solution tend to the corresponding values of $u^c$. When all the differences $u_j^k - u_j^c$ become smaller than $1/n$ the final relaxation of $\eta$ rapidly starts, with all the values of $\eta_j$ tending together to 0.5 and the dynamics stops. If one uses other approximations for the Heaviside function, as $\eta_2$ or $\eta_3$, the decay activation can be less sharp but the behavior remains essentially the same.

In analogy with what we said for the continuous problem, it follows also directly from (16) that (*) holds if at some iteration the vector $\eta^k$ is such that $\delta\eta_j^k \leq 0$ for any $j$ (a sort of discrete superharmonicity of $\eta^k$ at any node). This is not true in general from the beginning: for some iterations it could easily happen that the distance of the solution from the target increases at certain nodes, according to the data of the problem. This is the case for example if $u_j^0 = u_j^c$ at an internal node: the solution detaches from the target for a while before going back. Anyway, we could say that in a finite number of iterations, due to the boundary conditions, condition (*) is automatically satisfied.

We conclude this section with a couple of remarks. If the Heaviside under the Laplacian has the role of a switch, its approximated versions (6) or (7) do much than this: through their relaxed values in [0.5, 1] they are able to slow down the descent when the distance to the target tends to zero, avoiding jumps to subcritical values for the solution. Moreover, denoting by $t^*$ the time where the solution reaches the target up to a prescribed tolerance on the given grid, we have that when the parameter $n$ of the relaxed Heaviside function $\eta$ grows, then $t^*$ decreases. The closer is the $\eta$ to the real Heaviside function, the faster is the convergence to the target. Unfortunately there are stability limits for this process, as we will see in the tests, and $n$ cannot be chosen arbitrarily large.

# 4 Approximation on a Synchronized Family of Grids

In the previous section we have seen that on a given (fixed) grid the scheme (14) ((16) in the 1D case) is able to bring in time all the values of the initial solution towards the corresponding target values. If one desires great accuracy with respect to the target, then a large $m$ has to be chosen, with a consequent higher computational cost. Moreover, since the convergence progresses from the boundary towards the interior, one layer at a time, the entire process is heavily slowed down. That is why we tested a modified approach (partially inspired by Mosco [9]), based on an increasing family of grids $G_m$, with a precise synchronization of space and time steps which keeps constant the parameter $\gamma$.

The idea is the following. Scheme (14) is activated on an initial coarse grid $G_{m_0}$, for a given time interval $(0, T]$. For example, if $m_0 = 2$ and $T = 1$, we start to compute the solution with initial datum $u^*_{m_0} = u^0$ on 25 nodes, 9 of which internal, for 16 instant times (since on $G_2$ $h_2 = 1/2$ and $dt_2 = 1/16$).

When the scheme has been completed for all the instants of the first time interval the spatial grid is refined, from $G_{m_0}$ to $G_{m_1}$ (with $m_1 = m_0 + 1$), and we need to set a new initial value on it, the function $u^*_{m_1}$, in order to let the evolution start again on another time interval of length $T$. On the old nodes of $G_{m_0}$, which still belong to $G_{m_1}$, we simply use the values reached by the solution $u_{m_0}$ at the last time iteration, but on the added nodes (those in $G_{m_1} - G_{m_0}$) we have to introduce new values. One cannot make a direct use of the initial function $u_0$: since some time has passed we need a sort of actualization of it, and different strategies are possible for such an update process, keeping in mind that we are running in direction of the target. Then the scheme starts again with a reduced time step. In the previous example, after the refinement to $G_3$, the solution is computed on 81 nodes, 49 of which internal, for 64 instant times in $(1, 2]$ (since on $G_3$ now $h_3 = 1/4$ and $dt_3 = 1/64$). In such a way the process is repeated on a sequence of synchronized increasing grids, up to the desired finest grid $G_N$, where a suitable stopping criterion is imposed for the desired tolerance. In order to accelerate the process, one could introduce such criterion even at any previous grid, anticipating the refinements when the tolerance is achieved on the relative nodes.

The simplest update strategy for $u^*_m$ is to use a convex combination of the two initial data $u^0$ and $u^c$ with a suitable actualization coefficient $\lambda^m$:

$$(u^*_m)_{ij} = u^c_{ij} + \lambda^m (u^0_{ij} - u^c_{ij}); \tag{22}$$

we adopted for example $\lambda^m = 1/4^{m-m_0}$, in such a way that the values of $u^*_m$ tend towards those of $u^c$ as $m$ increases. Note that by this formula all the new values are naturally supercritical, property which is not guaranteed for example by the simple interpolation of the values of $u$ at the adjacent old nodes. But by using the same parameter $\lambda^m$ for any new node we do not take care of the different decay rates towards the target achieved on the previous grid. The consequence is that after the refinement some spurious oscillations can be introduced in the dynamics, at least for some iterations, before that the monotonicity behavior described in Proposition 3.1 comes back to act.

Such phenomenon can be reduced or even prevented by adopting other update strategies which make use of local values for $\lambda^m$. Here we limit ourselves to a couple of them in the simple 1D case.

In the first one we set for each new node $x_j$ of $G_m$:

$$(u_m^*)_j = u_j^c + \lambda_j^m (u_j^0 - u_j^c), \quad \lambda_j^m = \min_{s=j\pm 1} \frac{(u_{m-1}^{last})_s - u_s^c}{u_s^0 - u_s^c}, \tag{23}$$

where the quotient represents the decay rate of the solution at a node $x_s$ of $G_{m-1}$. In other words the best decay rate of the adjacent nodes is adopted.

Another idea is the following: in order to avoid oscillations after the update, we choose the values of $\eta$ at a new node $x_j$ in order to have a zero discrete Laplacian on it, that is

$$(u_m^*)_j = \eta^{-1}[(\eta_m^*)_j] + u_j^c, \quad \text{with} \quad (\eta_m^*)_j = \frac{(\eta_{m-1})_{j+1} + (\eta_{m-1})_{j-1}}{2}, \tag{24}$$

where $\eta^{-1}$ denotes the inverse function of $\eta$ in the range [0.5, 1]. With formula (24) the monotonicity property of Proposition 3.1 is preserved even after any refinement. As an example, when $\eta = \eta_3$ we easily get the explicit formula:

$$(u_m^*)_j = \frac{\tan(\pi((\eta_m^*)_j - 0.5))}{n} + u_j^c.$$

Summing up, the new scheme on the set of synchronized grids can be expressed by:

**Scheme (S2)** For $m = m_0, \ldots, N$

- compute $u_m^0 = \begin{cases} u_0 & \text{if } m = m_0 \\ u_m^* & \text{if } m > m_0 \end{cases}$.
- while $k \leq T/dt_m$ on $G_m$, for any $x_{ij} \in G_m$

$$\begin{cases} (u_m^{k+1})_{ij} = (u_m^k)_{ij} + dt_m \Delta_m (\eta_m^{k+1})_{ij} \\ \\ (u_m^{k+1})_{ij} = (u_m^k)_{ij} = u_{ij}^c \quad \text{at } x_{ij} \in \partial G_m \end{cases}. \tag{25}$$

- if $\|u_N^k - u^c\|_\infty = \max_{ij} |(u_N^k)_{ij} - u_{ij}^c| < tol$ on $G_N \Rightarrow$ STOP.

As usual $(\eta_m^k)_{ij} = \eta((u_m^k)_{ij} - u_{ij}^c)$ for the current active $m$, with $(\eta_m^k)_{ij} = 0.5$ on $\partial G_m$. The scheme then works on any grid for a fixed interval of time $T$, and $u_m^k$ represents the approximate solution on $G_m$ at time $t = (m - m_0)T + kdt_m$ (or even earlier if the stopping criterion is imposed on any grid).

We will see in the next section that Scheme (S2) usually speeds up the convergence towards the target, reducing at the same time the computational cost in a considerable way.

## 5 Numerical Tests

In this section we report the results of some numerical tests made applying schemes (S1) or (S2). We start with the simplified one dimensional case, in order to compare more easily the performance of the two approaches. Now $\Omega = (-1, 1)$, and problem (8) reduces to

$$u_t = (\eta(u - u^c))_{xx} \text{ in } \Omega \times (0, \infty), \quad u(0) = u^0 \text{ in } \Omega, \quad u = u^c \text{ on } \partial\Omega \times [0, \infty),$$

where $\eta$ denotes the approximated Heaviside function (one of those defined in Sect. 1). It means that for every internal node $x_i$ of $G_m$ we solve scheme (16), with, as usual, $\gamma = 1/4$. At any time step such nonlinear implicit system is solved via the Newton method (with a maximum number $itmax$ of iterations and a given tolerance $\tau$, usually $itmax = 5$ and $\tau = 10^{-5}$). We compare the behavior of solution of Scheme (S1) on the fixed grid $G_N$ with the one of Scheme (S2) on the increasing grid sequence $\{G_2, \ldots, G_N\}$ up to the same final accuracy: the computation ends when the distance between $u_N^k$ and $u^c$ in the infinity norm becomes less than a given tolerance. For (S2) it means in particular that the scheme will run on each mesh for a fixed interval of time or until the given tolerance is achieved, followed by a suitable mesh refinement and an upgrade process, up to the final grid $G_N$.

**TEST 1** $u^c(x) = 1 - |x|, \quad u^0(x) = 5(1 - |x|)$ [Fig. 1]

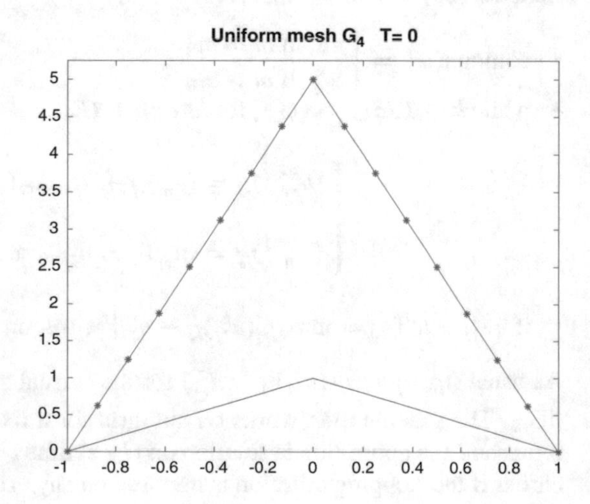


**Fig. 1** TEST 1. Initial datum (blue) and target function (red)

In Fig. 2 we compare the evolution of the solution of (S1) on $G_4$ with the one of (S2) on $\{G_2, G_3, G_4\}$ at the same time values: in both cases it progressively gets closer (decreasing) to $u^c$, from the boundary towards the interior. Since we adopted the function $\eta_1$ of (6) this essentially happens one node at a time. Each value of $u^0$ moves from its initial position only when the solution at an adjacent node becomes sufficiently close to the critical one. The convergence is monotone also in $\eta^k$ as stated by Proposition 3.1: its values, which at the beginning are equal to 1 at any internal node (since $u_0$ was largely supercritical), slowly relax towards the value 0.5 which characterizes the contact points. It means that the solution remains supercritical along the whole process. In (S2) a mesh refinement is applied after the same fixed time interval (in this case equal to one).

In Fig. 3 we show the evolution of the quantities $M_m^k$ and $E_m^k$; look in particular to the impulsive character produced by the specific choice of $\eta_1$ in Fig. 3b. In this test the assumptions of Proposition 3.1 are satisfied with $k_0 = 0$, and both the quantities decrease in a monotone way.

In Table 1 it is possible to compare the cost performances of the two schemes for this test: for any scheme we reported the stopping time $t^*$, the number of internal nodes involved for any grid, and the time iterations executed over each of them. This gives (in the last column) a total number of solution evaluations up to the stop. The stopping time is much shorter for the second scheme, in particular when the anisotropic update (23) is used, with an evaluation reduction of more than the 80% !

Finally in Fig. 4 we compare the different effects of the update formulas (22) and (23) on $u_3^*$ after the refinement from $G_2$ to $G_3$, showing the advantage of the second approach.

We remember that the stopping time does not depend on the specific choice of the approximated Heaviside function $\eta$, but only on the value of $\eta'(0)$, which in some sense determines the accuracy of the approximation. The following Table 2 shows, for equal values of $\eta'(0)$ in the three considered cases for $\eta$, how the stopping time on $G_4$ decreases as $n$ grows, in this case with a tolerance of $10^{-3}$. Since when $n$ grows $\eta$ approaches the real Heaviside function, one could imagine that $t^*$ would converge to the finite time of convergence of the continuous model (1). Unfortunately, due to stability problems, when $n$ becomes too large, the discrete solution can easily overcome the target in some nodes, causing oscillations and loops, so that the convergence gets lost.

**TEST 2** $u^c(x) = 1.5x^2(1 - x^2)\sin(1 + x), \quad u^0(x) = (1 - x^4)$ [Fig. 5a]

In this example the target is not symmetric, and the initial datum has no relation with it at all. Even in this case we have that the solution evolution with scheme (S1) is monotone (see Fig. 5b, c). Note the little different behavior induced by the chosen $\eta$ function, even if the final stopping time is essentially the same. In Fig. 6 we see how the scheme (S2) works in this case: the solution gets close to the target already on the coarsest grid, so that the refinements only act to extend the accuracy to the whole increasing set of nodes.

On this example we tested the time of convergence of the solution to the target, which of course depends on the chosen tolerance for the stopping criterion.

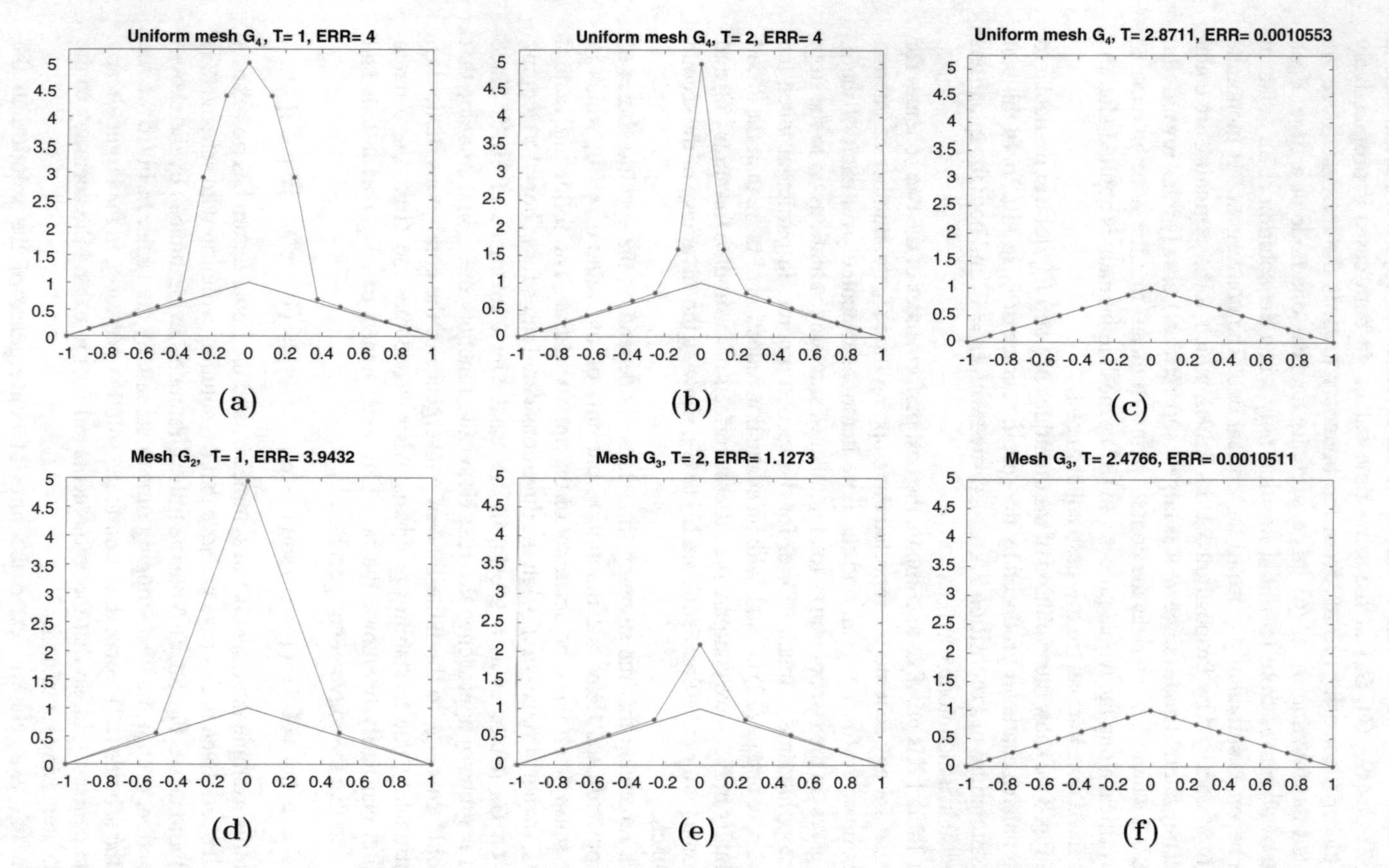


**Fig. 2** TEST 1. Evolution of $u$, (S1) versus (S2) ($\eta_1$, $tol = 10^{-3}$, update (22)). **(a)** S1: $u$ on $G_4$, $t = 1$. **(b)** S1: $u$ on $G_4$, $t = 2$. **(c)** S1: $u$ on $G_4$, $t = t_1^*$. **(d)** S2: $u_2$ on $G_2$, $t = 1$. **(e)** S2: $u_3$ on $G_3$, $t = 2$. **(f)** S2: $u_4$ on $G_4$, $t = t_2^* < t_1^*$

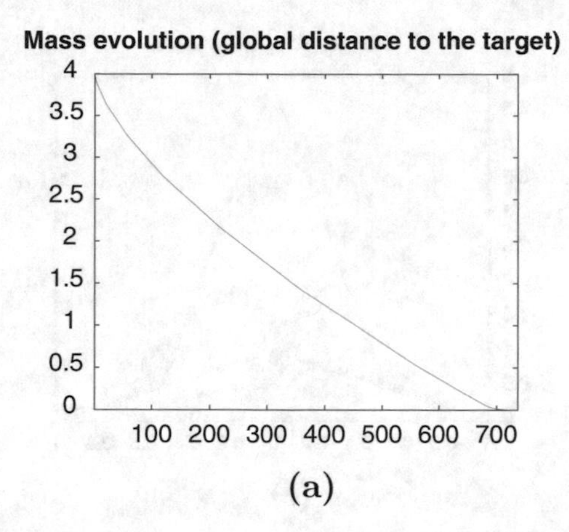


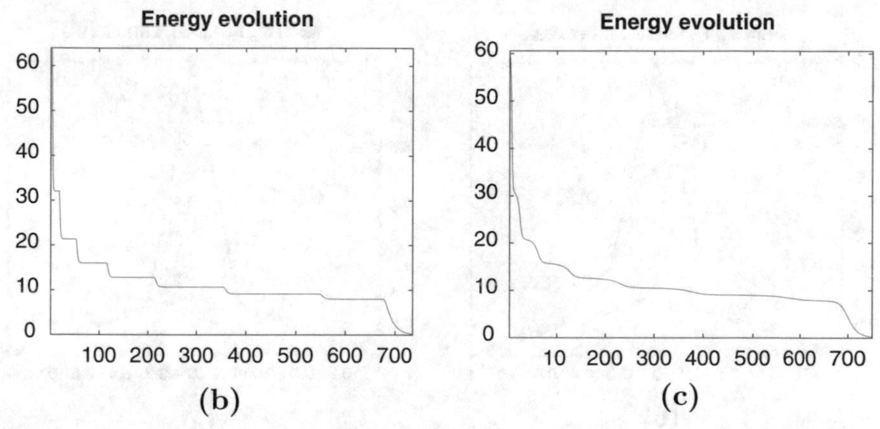


**Fig. 3** TEST 1. $M^k$ and $E^k$ evolution with (S1) on $G_4$. (**a**) $M^k$. (**b**) $E^k$ with $\eta_1$. (**c**) $E^k$ with $\eta_3$

**Table 1** TEST 1. Performance comparison of schemes (S1) and (S2)

| Scheme | $t^*$ | Grid | Nodes | Time it. | Evals | Tot. evals |
|---|---|---|---|---|---|---|
| S1 | 2.8711 | $G_4$ | 15 | 736 | 11,040 | 11,040 |
| S2 | | $G_2$ | 3 | 16 | 48 | |
| S2 | | $G_3$ | 7 | 64 | 448 | |
| S2 + (22) | 2.4766 | $G_4$ | 15 | 123 | 1845 | 2341 |
| S2 + (23) | 2.2695 | $G_4$ | 15 | 70 | 1050 | 1546 |

Experiments suggest that its rate is exponential: Table 3a below shows that there exist two positive constants $C$ and $\alpha$ such that

$$\|u(t) - u^c\|_\infty \simeq Ce^{-\alpha t}.$$

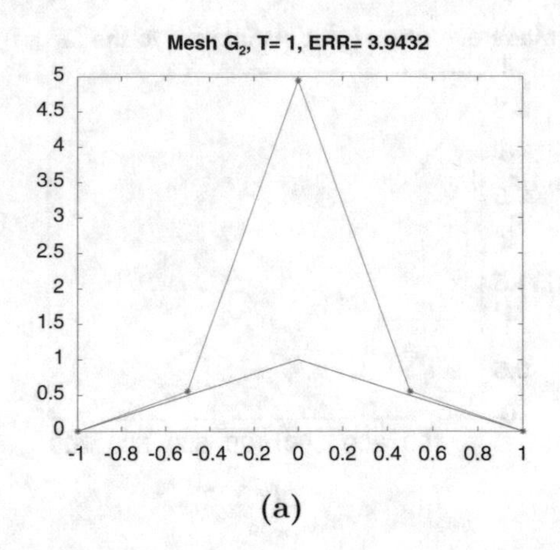


(a)

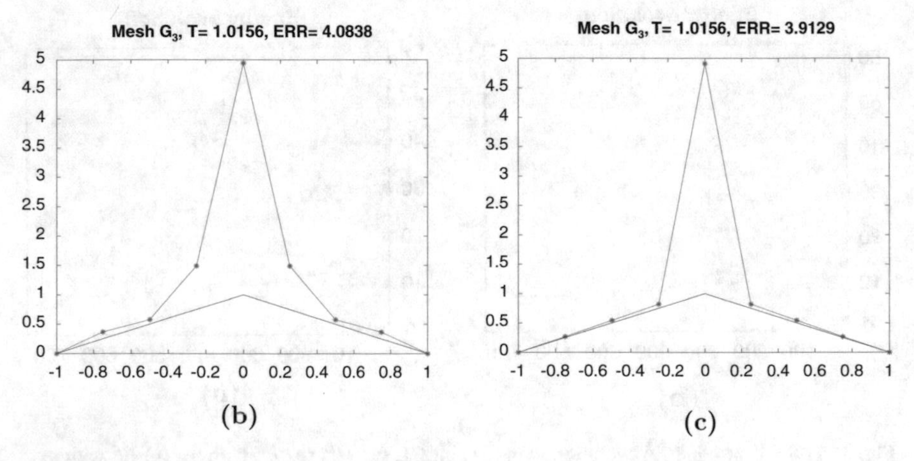


(b) (c)

**Fig. 4** TEST 1. Scheme (S2): the update process after $G_2 - G_3$ refinement. (**a**) $u_2$ on $G_2,\ t = 1$. (**b**) $u_3^*$ with (22). (**c**) $u_3^*$ with (23)

**Table 2** TEST 1. Stopping time $t^*$ for different $\eta$ functions on $G_4$ when $n$ grows

| $\eta'(0)$ | $\eta_1$ | $\eta_2$ | $\eta_3$ |
|---|---|---|---|
| 7.5 | 2.8711 | 2.8789 | 2.918 |
| 15 | 2.7656 | 2.7656 | 2.7891 |
| 30 | 2.7188 | 2.7188 | 2.7305 |

We could say that in the discrete case the finite time of convergence to the target is the one corresponding to the machine precision (of order $10^{-16}$). On the contrary, we see that the final time does not grow with the chosen order of the grid $G_N$ of interest (Table 3b).

**TEST 3** $u^c(x) = 1 - |x|, \quad u^0$ given vector [Fig. 7a]

In this case the initial datum is a vector of values on $G_4$ very close to the target at some nodes, and even equal to that at $x = 0$. Looking at the graph of $\eta^0$ (Fig. 7e) we

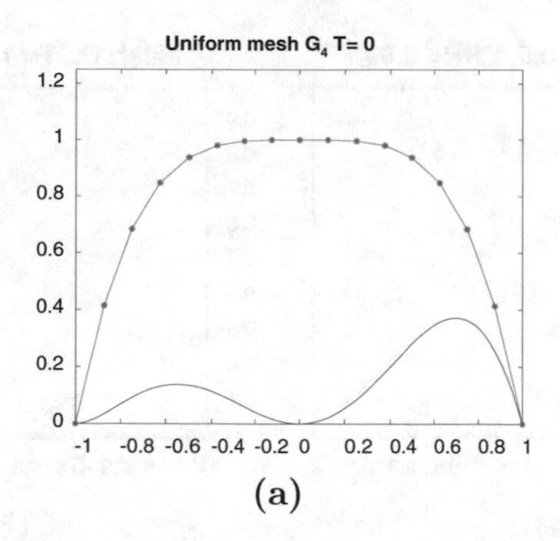


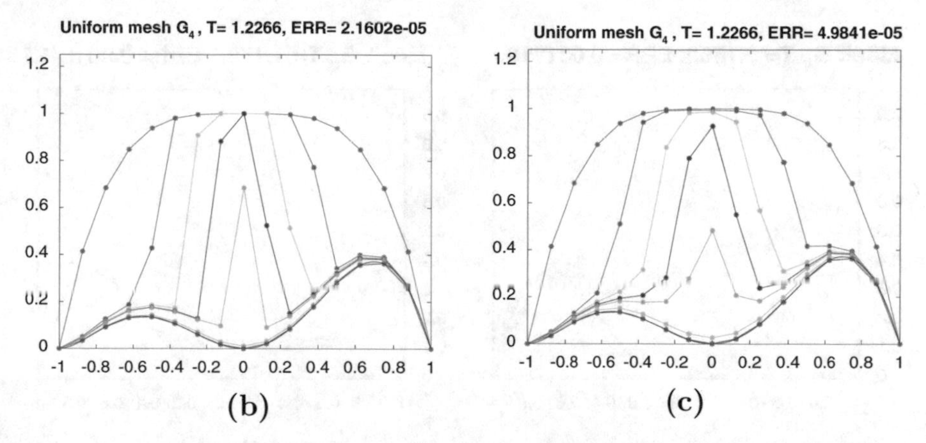


**Fig. 5** TEST 2. (in b) and (c) the solution of (S1) is plotted every 45 iterations. (**a**) Test 2: initial data. (**b**) (S1) on $G_4$ with $\eta_1$. (**c**) (S1) on $G_4$ with $\eta_3$

can see that $\delta\eta_j^0 > 0$ at some nodes: there the solution initially grows (in particular it detaches from the target at the center). But after a few iterations $\delta\eta_j^k \leq 0$ at any node (Fig. 7h) and it starts to decrease everywhere, entering in the "monotone regime." This behavior is confirmed from the corresponding evolution of $E^k$ (Fig. 8) which starts decreasing only after some iterations.

**TEST 4** $u^c(x) = 1 - |x|, \quad u^0 = 5(1 - |x|) + 1 + x \quad$ [Fig. 9]

In this example the contact between $u^0$ and $u^c$ is only at the left boundary point. Nevertheless this is enough to produce the transmission of information up to the opposite side, so that all the solution is again absorbed by the target. It is sufficient to impose a homogeneous Neumann boundary condition on the opposite side (see Fig. 9). In other words what is requested in order to start the global process is a boundary contact of the initial datum with the target.

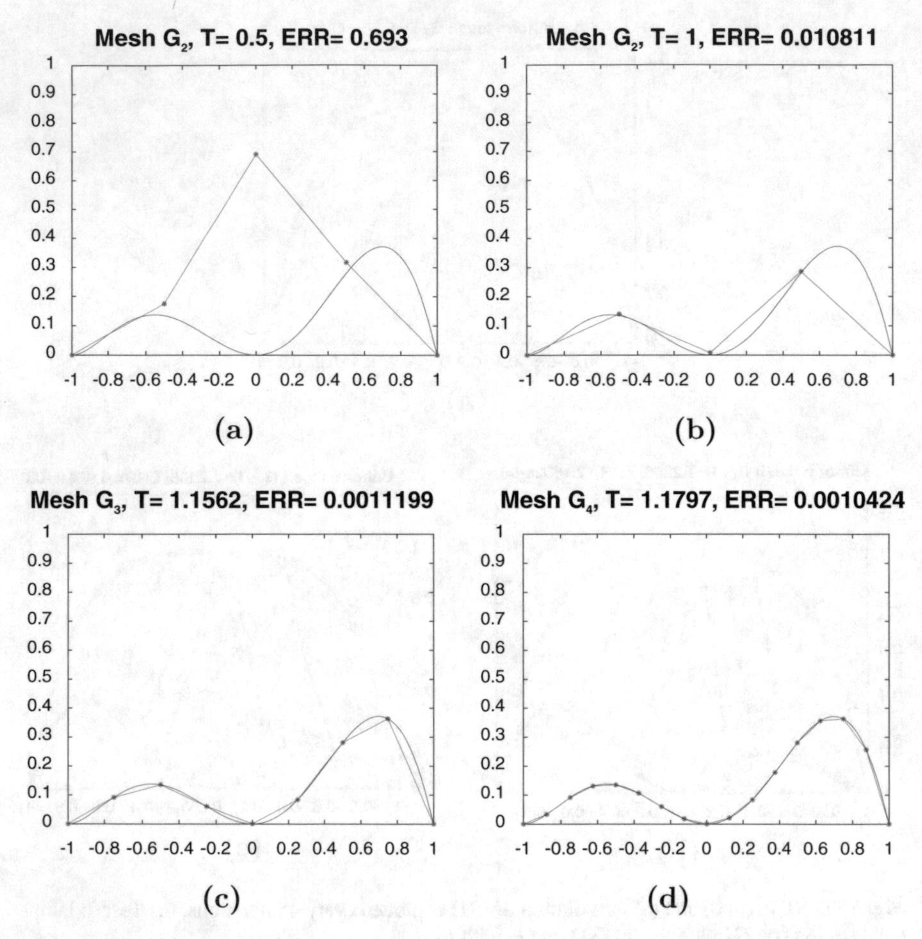


**Fig. 6** TEST 2. Solution evolution with scheme (S2). (**a**) $G_2,\ t = 0.5$. (**b**) $G_2,\ t = 1$. (**c**) $G_3,\ t = 1.15$. (**d**) $G_4,\ t = 1.175$

**Table 3** TEST 2. ($\eta_1,\ n = 10$) stopping time for (S1) versus (a) tolerance, (b) grid order

| $tol$ | Stopping time (on $G_4$) | $grid$ | Stopping time ($tol = 10^{-5}$) |
|---|---|---|---|
| $10^{-4}$ | 1.1367 | 3 | 1.3281 |
| $10^{-8}$ | 1.6562 | 4 | 1.2695 |
| $10^{-12}$ | 2.1719 | 5 | 1.2539 |
| $10^{-16}$ | 2.7007 | 6 | 1.2507 |

**TEST 5** $u^c(x, y) = 1 - \frac{|x+y|+|y-x|}{2}, \quad u^0(x, y) = 5\left(1 - \frac{|x+y|+|y-x|}{2}\right)$ [Fig. 10a]

We end with a test on the more general 2D case, where all the previous behaviors are confirmed. We generalized the initial data choice of TEST 1 assuming as

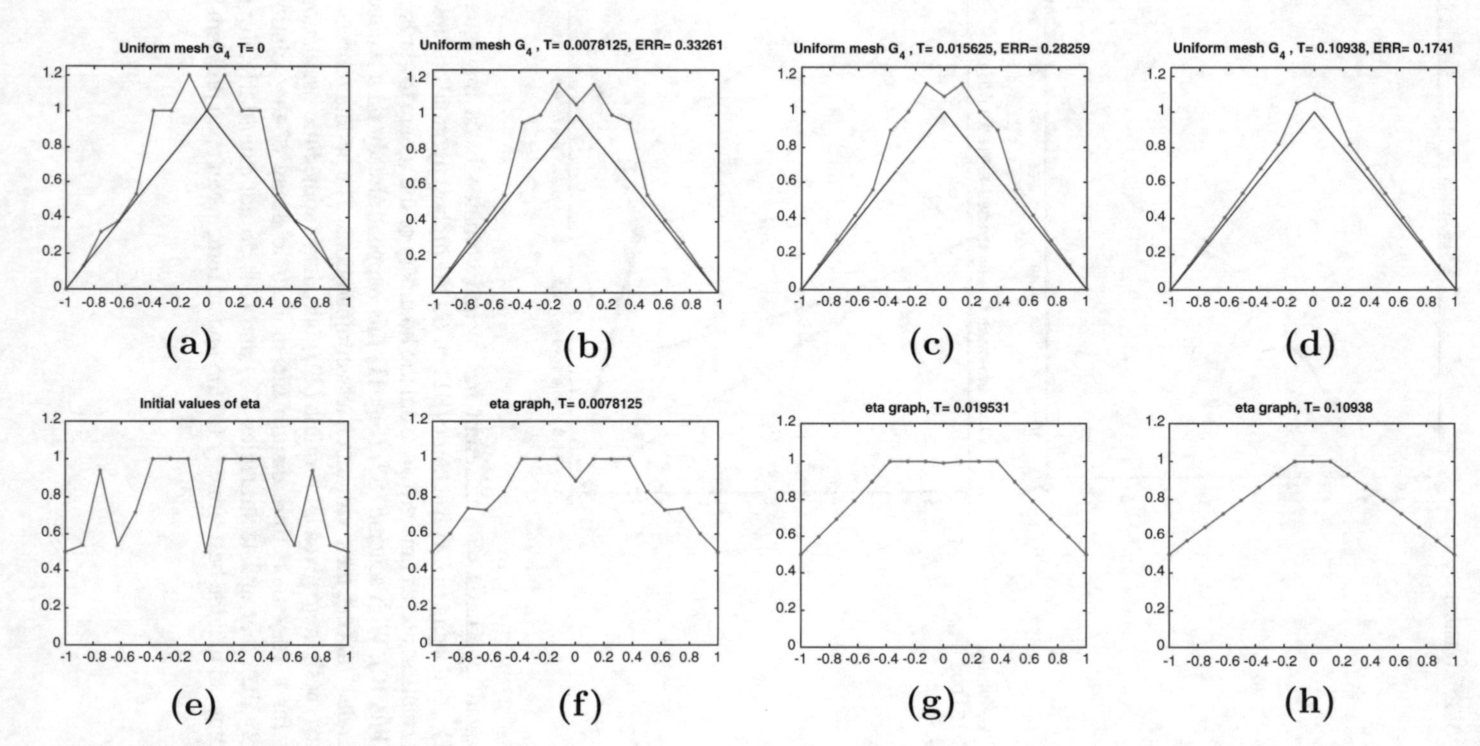


**Fig. 7** Test 3. Scheme (S1), $u$ versus $\eta$. **(a)** initial data on $G_4$. **(b)** $u$ at $it = 1$. **(c)** $u$ at $it = 4$. **(d)** $u$ at $it = 27$. **(e)** initial $\eta$ on $G_4$. **(f)** $\eta$ at $it = 1$. **(g)** $\eta$ at $it = 4$. **(h)** $\eta$ at $it = 27$

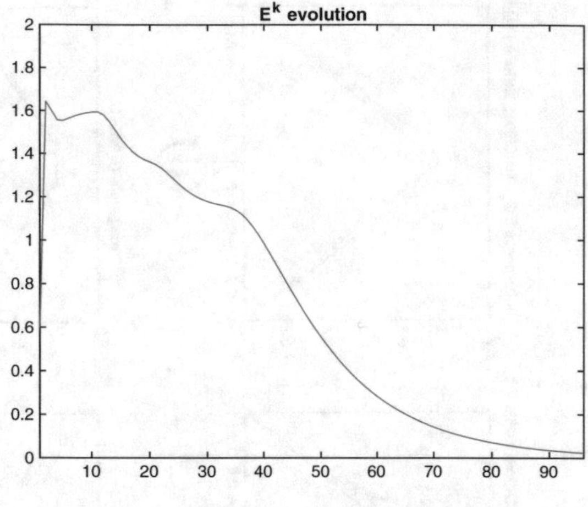


**Fig. 8** TEST 3. $E^k$ evolution

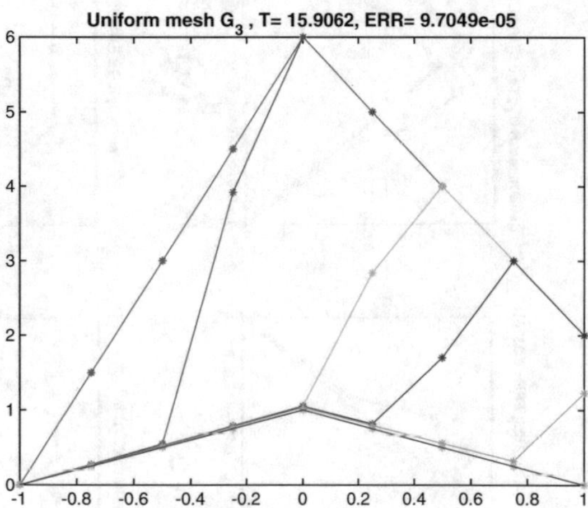


**Fig. 9** TEST 4. Solution evolution with (S1) on $G_3$, times $t = 1, 6, 9, 13$

data two square pyramids with different height and common basis the square $\Omega = (-1, 1)^2$, and using $\eta_1$. With scheme (S1) on $G_4$ the highest pyramid collapses towards the smallest one as time grows, from the boundary of the square towards its center (see Fig. 10). With scheme (S2) (Fig. 11) this happens already on a reduced number of nodes (on $G_3$), then with much less computations. The update process at any change of mesh (here given by formula (22)) introduces some little oscillations which only for a very short time disturb the monotonic decay of the solution. Unfortunately the other update strategies (23) and (24) do not extend trivially to the 2D settings, so that the best strategy for the update remains an open problem in this case.

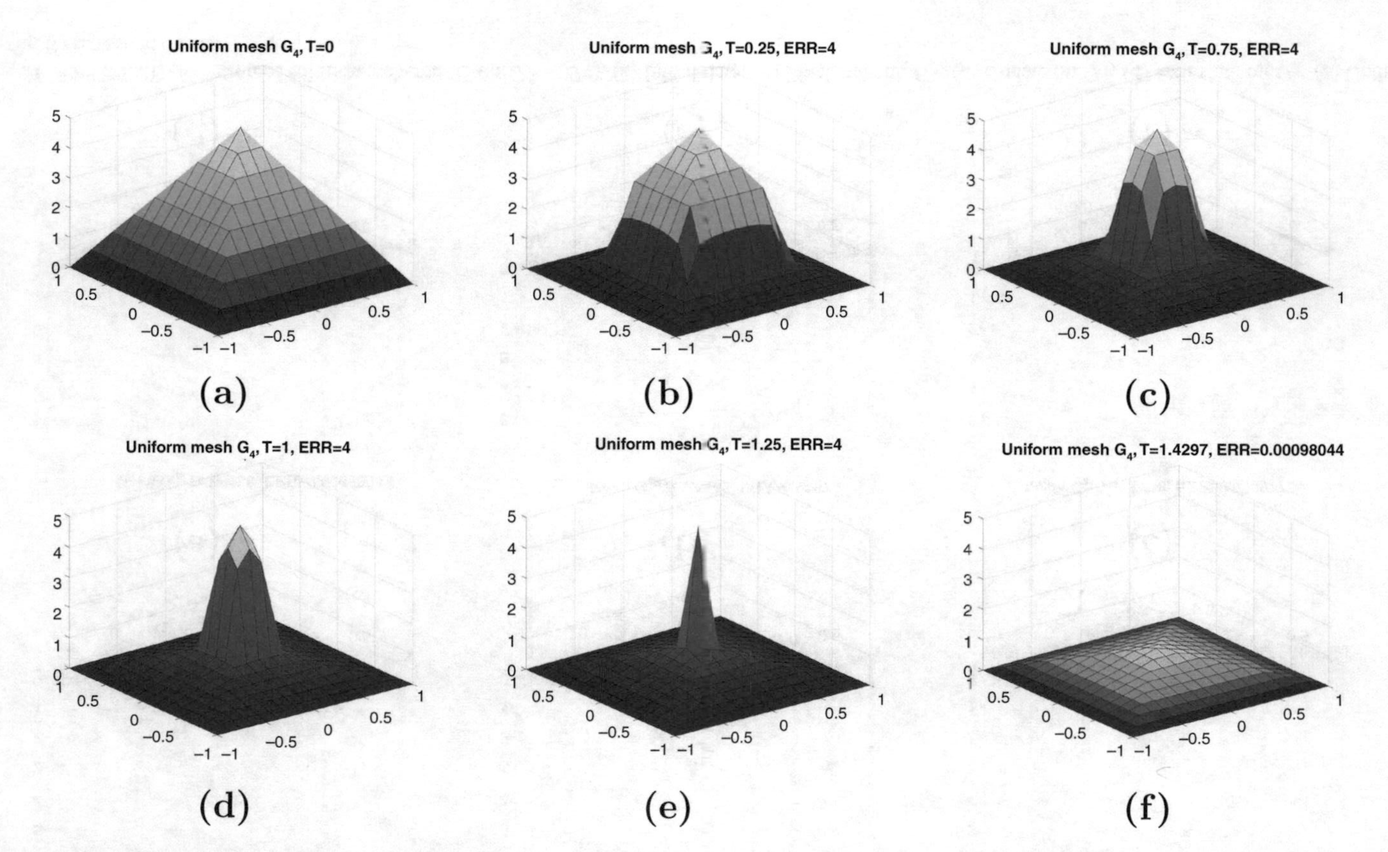


**Fig. 10** TEST 5. 2D-evolution of solution (scheme S1 on $G_4$). **(a)** Initial data. **(b)** Time $t = 0.25$. **(c)** Time $t = 0.75$. **(d)** Time $t = 1$. **(e)** Time $t = 1.25$. **(f)** Exit state

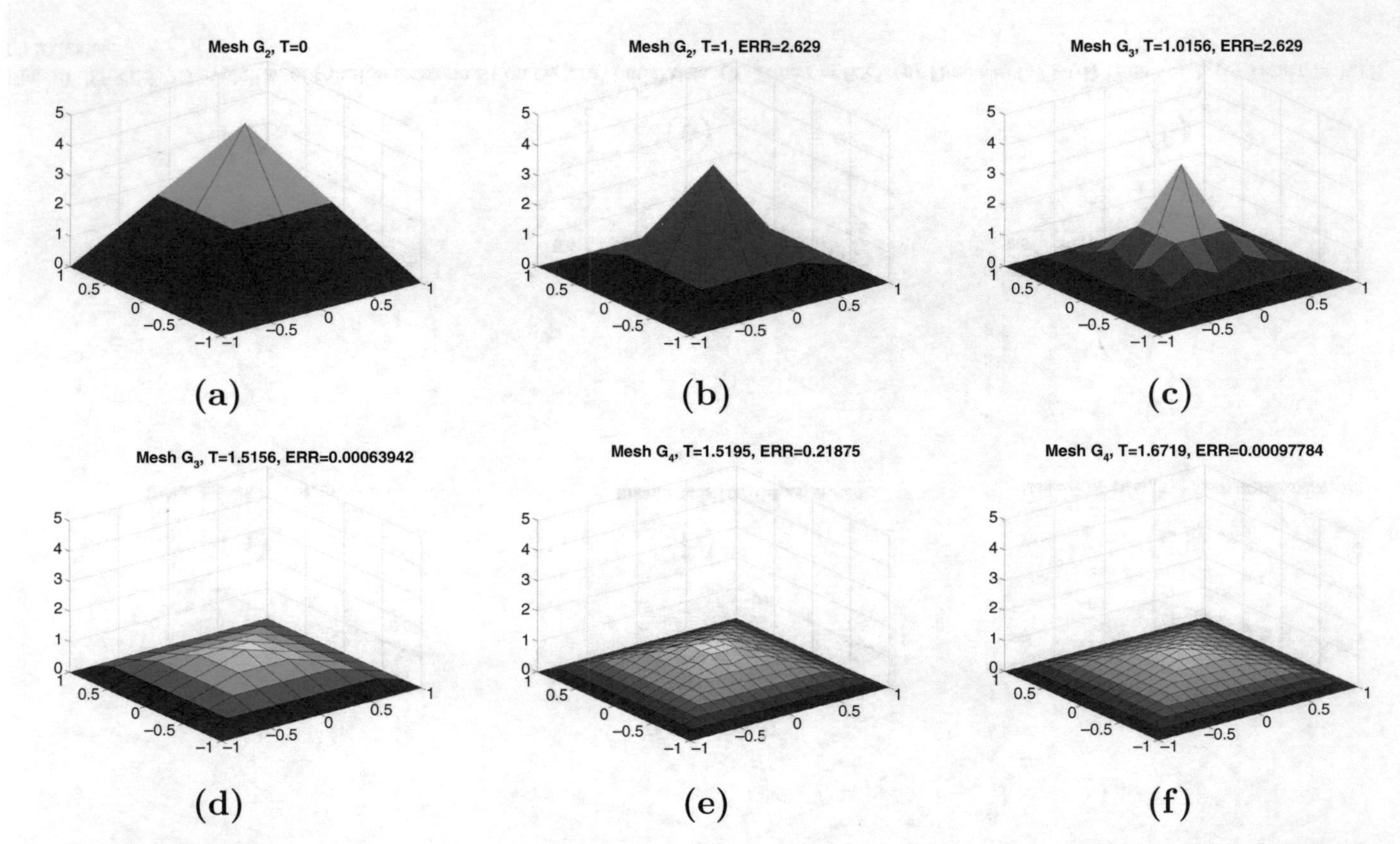


**Fig. 11** TEST 5. 2D-evolution of solution (scheme S2 on $G2 - G4$). **(a)** Initial data. **(b)** Final state on $G_2$. **(c)** Update on $G_3$. **(d)** Exit state on $G_3$. **(e)** Update on $G_4$. **(f)** Exit state on $G_4$

**Acknowledgments** We wish to thank Umberto Mosco and Maria Agostina Vivaldi for their helpful support and fruitful discussions while working on this paper.

## References

1. Bak P., Tang C., Wiesenfeld, K.: Self-organized criticality: an explanation of the 1/f noise. Phys. Rev. Lett. **59**, 381–394 (1987)
2. Bak, P., Tang, C., Wiesenfeld, K.: Self-organized criticality. Phys. Rev. A **38**(3), 364–374 (1988)
3. Bántay, P., Jánosi, M.: Self-organization and anomalous diffusion. Phys. A **185**, 11–18 (1992)
4. Barbu, V.: Self-organized criticality and convergence to equilibrium of solutions to nonlinear diffusion equations. Annu. Rev. Control **34**, 52–61 (2010)
5. Barbu, V.: Self-organized criticality of cellular automata model; absorbtion in finite-time of supercritical region into the critical one. Math. Methods Appl. Sci. **36**, 1726–1733 (2013)
6. Carlson, J.M., Swindle, G.H.: Self-organized criticality: sandpiles, singularities, and scaling. Proc. Natl. Acad. Sci. USA **92**, 6712–6719 (1995)
7. Dahr, D.: Self-organized critical state of sandpile automaton models. Phys. Rev. Lett. **64**, 1613 (1990)
8. Ion, S., Marinoschi, G.: A self-organizing criticality mathematical model for contamination and epidemic spreading. Discrete Continuous Dynam. Syst. B **22**(2), 383–405 (2017)
9. Mosco, U.: Finite-time self-organized-criticality on synchronized infinite grids. SIAM J. Math. Anal. **50**(3), 2409–2440 (2018)

# Approximation of 3D Stokes Flows in Fractal Domains

**Massimo Cefalo, Simone Creo, Mirko Gallo, Maria Rosaria Lancia, and Paola Vernole**

**Abstract** We study a Stokes flow in a cylindrical-type fractal domain with homogeneous Dirichlet boundary conditions. We consider its numerical approximation by mixed methods: finite elements in space and finite differences in time. We introduce a suitably refined mesh à la Grisvard, which in turn will allow us to obtain an optimal a priori error estimate.

**Keywords** Stokes equation · Fractals · Finite elements · Weighted Sobolev spaces

## 1 Introduction

In this paper we study an incompressible Stokes equation, coupled with no-slip boundary conditions, in a three-dimensional cylindrical domain, which is the cartesian product between a two-dimensional pre-fractal domain of Koch type and the unit interval. This cylindrical set, which is polyhedral and non-convex, can model a rough micro-channel.

Nowadays, many papers deal with the study of viscous flows in rough micro-channels; this is due to the fast development of the so-called MEMS (which stands

---

M. Cefalo
Dipartimento di Ingegneria Informatica, Automatica e Gestionale, Sapienza Università di Roma, Roma, Italy

S. Creo (✉) · M. R. Lancia · P. Vernole
Dipartimento di Scienze di Base e Applicate per l'Ingegneria, Sapienza Università di Roma, Roma, Italy
e-mail: simone.creo@sbai.uniroma1.it; mariarosaria.lancia@sbai.uniroma1.it; vernole@mat.uniroma1.it

M. Gallo
Dipartimento di Ingegneria Meccanica e Aerospaziale, Sapienza Università di Roma, Roma, Italy
e-mail: mirko.gallo@uniroma1.it



M. R. Lancia, A. Rozanova-Pierrat (eds.), *Fractals in Engineering: Theoretical Aspects and Numerical Approximations*, SEMA SIMAI Springer Series 8,
https://doi.org/10.1007/978-3-030-61803-2_2

for Micro Electro-Mechanical Systems), e.g. micro-motors micro-turbines. Usually, Stokes equations model these phenomena.

The characteristics of the flow deeply affect the design and the process control of MEMS and heat transfer processes [31, 35, 37]. Fractal-type structures in the study of Stokes problems can be of great interest from the point of view of applications, firstly because they can model rough and irregular geometries, secondly because they could be important in the design of microfluidic devices. In fact, the mechanical stress exerted by the fluid onto solid boundaries is significant, when a large value of the velocity gradient towards the wall is achieved [33]. In this context, fractal-type geometries could mitigate these effects.

The study of Stokes problems in irregular domains is part of a long term project: the study of vector BVPs in fractal domains. This topic is rather recent and it is fast developing; for some literature on vector BVPs in irregular domains, we refer to [1, 8, 12, 20, 21, 26]. As to the study of scalar BVPs in fractal-type domains, among the others we refer to [9–11, 13, 25, 27].

The main focus of this paper is on the numerical approximation of Stokes problems in fractal-type structures and on the proof of optimal a priori error estimates both in space and in time. We point out that the irregular nature of fractal-type boundaries does not allow us to use the standard techniques because the presence of irregular geometries deteriorates the regularity of the solution. For the numerical approximation of boundary value problems in fractal domains we refer to [5–7].

More precisely, for every $n \in \mathbb{N}$, we consider parabolic incompressible Stokes problems $(\tilde{P}_n)$ in a cylindrical domain $Q_n$ with a Koch-type cross section (see Sect. 2), with homogeneous Dirichlet boundary conditions:

$$(\tilde{P}_n)\begin{cases} \frac{\partial \mathbf{u}_n}{\partial t}(t,x) - \Delta \mathbf{u}_n(t,x) + \nabla p_n(t,x) = \mathbf{f}_n(t,x) & \text{in } [0,T]\times Q_n,\\ \operatorname{div}\mathbf{u}_n(t,x) = 0 & \text{on } [0,T]\times Q_n,\\ \mathbf{u}_n(t,x) = 0 & \text{on } [0,T]\times \partial Q_n,\\ \mathbf{u}_n(0,x) = \mathbf{u}_n^0(x) & \text{in } Q_n. \end{cases}$$

Existence and uniqueness results for the weak solution $\mathbf{u}_n$ of $(\tilde{P}_n)$ and for its associated pressure $p_n$, for every $n \in \mathbb{N}$, are provided in [26], along with the corresponding results in the limit fractal domain $Q$ (obtained in the limit as $n \to +\infty$). Moreover, it is proved that the couple $(\mathbf{u}, p)$, given by the solution of the Stokes problem in the fractal domain and by its associated pressure, can be suitably approximated in terms of $(\mathbf{u}_n, p_n)$.

The numerical approximation of problem $(\tilde{P}_n)$ is carried out by mixed methods, namely finite elements in space and finite differences in time. Since $Q_n$ is a non-convex domain, in order to prove an optimal a priori error estimate for the semi-discrete error, we construct a suitable mesh à la Grisvard [17], thus extending to the 3D case the results of [4]. We adapt to our case the regularity results in weighted Sobolev spaces for the couple $(\mathbf{u}_n, p_n)$ provided in [3], where a non-conforming method on pentahedral and tetrahedral meshes is obtained by using Crouzeix – Raviart interpolation operators [14].

We then discretize the problem in time by an implicit method. We obtain optimal a priori error estimates by means of suitable discrete Sobolev-type norms (see [18, 19]). We conclude by presenting some numerical simulations. Moreover, we prove that the average wall shear stress vanishes as $n \to +\infty$ for Hagen–Poiseuille-like flows.

The paper is organized as follows. In Sect. 2 we introduce the functional setting of this paper. In Sect. 3 we recall the existence and uniqueness results for the solution of problem $(P_n)$ and its associated pressure and we introduce two equivalent weak formulations, which allow us to give an "efficient" discrete version of the space of divergence-free functions. In Sect. 4 we introduce the weighted Sobolev spaces and we recall the regularity results for the couple $(\mathbf{u}_n, p_n)$. In Sect. 5 we prove that the average wall shear stress tends to zero as $n \to +\infty$. In Sect. 6 we approximate problem $(\tilde{P}_n)$ by FEM in space and FD in time and we prove optimal a priori error estimates. In Sect. 7 we present the numerical simulations of the problem at hand.

## 2 Preliminaries

We denote by $|P - P_0|$ the Euclidean distance between two points in $\mathbb{R}^3$. By the Koch snowflake $K$ we denote the union of three coplanar Koch curves $K_i$ (see [16]). We point out that $K$ has Hausdorff dimension $d_f = \frac{\ln 4}{\ln 3}$.

We assume that the junction points $A_1$, $A_3$, and $A_5$ are the vertices of a regular triangle with unit side length, i.e. $|A_1 - A_3| = |A_1 - A_5| = |A_3 - A_5| = 1$. One can define, in a natural way, a finite Borel measure $\mu_K$ supported on $K$ by

$$\mu_K := \mu_1 + \mu_2 + \mu_3,$$

where $\mu_i$ denotes the normalized $d_f$-dimensional Hausdorff measure restricted to $K_i$, for $i = 1, 2, 3$.

We denote by $K_n$ the pre-fractal closed polygonal curve approximating $K$ at the $n$-th step. We define $S_n = K_n \times I$, with $I = [0, 1]$; $S_n$ is a surface of polyhedral type. We give to a point $P \in S_n$ the Cartesian coordinates $P = (x, x_3)$, where $x = (x_1, x_2)$ are the coordinates of the orthogonal projection of $P$ on the plane containing $K_n$ and $x_3$ is the coordinate of the orthogonal projection of $P$ on the $x_3$-line containing the interval $I$.

By $\Omega_n \subset \mathbb{R}^2$ we denote the open bounded two-dimensional domain with boundary $K_n$. By $Q_n = \Omega_n \times I$ we denote the cylindrical-type domain having $S_n$ as lateral surface and the sets $\Omega_n \times \{0\}$ and $\Omega_n \times \{1\}$ as bases.

The measure on $S_n$ is

$$\mathrm{d}\sigma = \mathrm{d}\ell \times \mathrm{d}x_3,$$

where $d\ell$ is the arc-length measure on $K_n$ (which can be naturally defined) and $dx_3$ is the one-dimensional Lebesgue measure on $I$.

We introduce the fractal surface $S = K \times I$ given by the Cartesian product between $K$ and $I$. On $S$ we can define the finite Borel measure

$$dg = d\mu_K \times dx_3.$$

We remark that $S$ has Hausdorff dimension $d_f + 1$.

By $\Omega \subset \mathbb{R}^2$ we denote the two-dimensional fractal domain whose boundary is $K$. By $Q$ we denote the open cylindrical domain where $S = K \times I$ is the lateral surface and the sets $\Omega \times \{0\}$ and $\Omega \times \{1\}$ are the bases, see Fig. 1.

We point out that the sequence $\{Q_n\}$ is an increasing sequence invading $Q$, i.e.

$$\mathcal{L}(Q \setminus Q_n) \xrightarrow[n\to+\infty]{} 0,$$

where $\mathcal{L}$ is the Lebesgue measure in $\mathbb{R}^3$.

We now recall a trace theorem specialized to our case. For the proof we refer to Theorem 1 of Chapter VII in [23].

**Theorem 2.1** *Let $\Gamma$ denote $\Omega \times \{0\}$ and $\Omega \times \{1\}$. Then, for every $s > \frac{1}{2}$, $H^{s-\frac{1}{2}}(\Gamma)$ is the trace space of $H^s(Q)$, i.e.:*

1. *there exists a linear and continuous operator $\gamma_0 \colon H^s(Q) \to H^{s-\frac{1}{2}}(\Gamma)$;*
2. *there exists a linear and continuous operator* Ext: $H^{s-\frac{1}{2}}(\Gamma) \to H^s(Q)$, *such that $\gamma_0 \circ$ Ext is the identity operator on $H^{s-\frac{1}{2}}(\Gamma)$.*

Let $T$ be a compact set of $\mathbb{R}^N$. By $C(T)$ we denote the space of continuous functions on $T$ and by $C_0^\infty(T)$ the space of continuous infinitely differentiable functions with compact support in $T$.

Let $M$ be an open set of $\mathbb{R}^3$. By $L^2(M)$ we denote the Lebesgue space with respect to the Lebesgue measure $\mathcal{L}$. By $H^1(M)$ we denote the usual Sobolev space

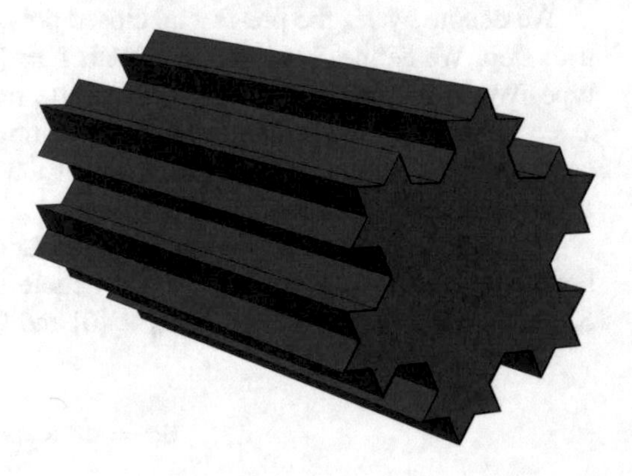

**Fig. 1** The fractal domain $Q$

and by $\|\cdot\|_{H^1(M)}$ we denote its norm. We set $H_0^1(M) = \overline{C_0^\infty(M)}^{\|\cdot\|_{H^1(M)}}$ and we denote by $H^{-1}(M)$ its dual.

In the following we define $H_0^1(Q_n)$ as the closure in $H^1(Q)$ of all functions $\phi \in C_0^\infty(Q)$ such that $\phi = 0$ on $Q \setminus Q_n$.

We define $L^2(M)^3 = \{\mathbf{u} = (u_1, u_2, u_3) : u_i \in L^2(M)\}$ and $L^2_\sigma(M) = \overline{(C_{0,\sigma}^\infty(M))}^{L^2(M)^3}$, where $C_{0,\sigma}^\infty(M) = \{\mathbf{v} \in C_0^\infty(M)^3 : \operatorname{div} \mathbf{v} = 0\}$; we endow $L^2_\sigma(M)$ with the $L^2$-scalar product.

We introduce the space (see [32, page 25]):

$$C_0^\infty([0,T); C_{0,\sigma}^\infty(M)) := \left\{ \mathbf{u}|_{[0,T)\times M} : \mathbf{u} \in C_{0,\sigma}^\infty((-1,T)\times M)^3, \operatorname{div} \mathbf{u} = 0 \right\}.$$

By $H_{0,\sigma}^1(M) = \overline{(C_{0,\sigma}^\infty(M))}^{H^1(M)}$ we denote the closed subspace of $H_0^1(M)^3$.

Moreover the following result holds (see Proposition 2.2. in [28]).

**Proposition 2.2** *The space $H_{0,\sigma}^1(M)$ is dense in $H_0^1(M)^3$.*

We denote by $\mathcal{P}\colon L^2(M)^3 \to L^2_\sigma(M)$ the Leray–Helmholtz projection and by $J$ the canonical injection $L^2_\sigma(M) \hookrightarrow L^2(M)^3$. We have that the adjoint of $J$ is $J' = \mathcal{P}$ and that $\mathcal{P} \circ J$ is the identity on $L^2_\sigma(M)$. The canonical injection $\tilde{J}\colon H_{0,\sigma}^1(M) \hookrightarrow H_0^1(M)^3$ is the restriction of $J$ to $H_{0,\sigma}^1(M)$. We denote by $\tilde{\mathcal{P}}$ the adjoint of $\tilde{J}$; since $\tilde{J}$ is the restriction of $J$ to $H_{0,\sigma}^1(M)$, $\tilde{\mathcal{P}}$ is an extension of $\mathcal{P}$ to $(H_{0,\sigma}^1(M))'$ (the dual of $H_{0,\sigma}^1(M)$). For more details on the Leray–Helmholtz projection operator see [32, Chapter II, Section 2.5].

## 3 Existence and Uniqueness Results

We now recall the main results about the Stokes problem in a pre-fractal domain, see [26].

The problem is formally stated as follows:

$$(\tilde{P}_n)\begin{cases} \frac{\partial \mathbf{u}_n}{\partial t}(t,x) - \Delta \mathbf{u}_n(t,x) + \nabla p_n(t,x) = \mathbf{f}_n(t,x) & \text{in } [0,T]\times Q_n, \\ \operatorname{div} \mathbf{u}_n(t,x) = 0 & \text{on } [0,T]\times Q_n, \\ \mathbf{u}_n(t,x) = 0 & \text{on } [0,T]\times \partial Q_n, \\ \mathbf{u}_n(0,x) = \mathbf{u}_n^0(x) & \text{in } Q_n. \end{cases}$$

We define for every $n \in \mathbb{N}$, the bilinear symmetric form $a_n(\mathbf{u}, \mathbf{v}) : H_{0,\sigma}^1(Q_n) \times H_{0,\sigma}^1(Q_n) \to \mathbb{R}$:

$$a_n(\mathbf{u}, \mathbf{v}) = \int_{Q_n} \nabla \mathbf{u} \cdot \nabla \mathbf{v} \, d\mathcal{L}. \tag{3.1}$$

We note that it is coercive in $H^1_{0,\sigma}(Q_n)$ thanks to Poincaré inequality and it is also closed in $L^2_\sigma(Q_n)$.

By Kato's theorem [24], there exists a unique non-positive self-adjoint operator $A^n_0\colon H^1_{0,\sigma}(Q_n) \to (H^1_{0,\sigma}(Q_n))'$ such that

$$a_n(\mathbf{u}, \mathbf{v}) = -\langle A^n_0 \mathbf{u}, \mathbf{v}\rangle_{(H^1_{0,\sigma}(Q_n))', H^1_{0,\sigma}(Q_n)}. \tag{3.2}$$

Moreover, the following holds (see [28]).

**Proposition 3.1** *Let $\Delta\colon H^1_0(Q_n) \to H^{-1}(Q_n)$ denote the Dirichlet Laplacian on $Q_n$. There holds*

$$A^n_0 = \tilde{\mathcal{P}} \circ (\Delta)\tilde{J}.$$

We call *the Stokes operator* $-A_n$ the part of $-A^n_0$ in $L^2_\sigma(Q_n)$, i.e. $D(A_n) = \{\mathbf{u} \in H^1_{0,\sigma}(Q_n)) \, : \, A^n_0 \mathbf{u} \in L^2_\sigma(Q_n)\}$ and $A_n \mathbf{u} = A^n_0 \mathbf{u}$.

**Theorem 3.2** *The operator $A_n$ is self-adjoint in $L^2_\sigma(Q_n)$ and generates an analytic contraction semigroup $T_n(t)\colon L^2_\sigma(Q_n) \to D(A_n)$ with*

$$D(A_n) = \{\mathbf{u} \in H^1_{0,\sigma}(Q_n) \, : \, \exists\, p \in (D(Q_n))' \text{ s.t. } \nabla p \in H^{-1}(Q_n), -\Delta \mathbf{u} + \nabla p \in L^2_\sigma(Q_n)\}$$

*and*

$$-A_n \mathbf{u} = -\Delta \mathbf{u} + \nabla p \in L^2_\sigma(Q_n).$$

Moreover, from [32, Chapter 3, Lemma 2.2.1], there exists a unique positive self-adjoint operator $(-A_n)^{\frac{1}{2}}\colon D((-A_n)^{\frac{1}{2}}) \to L^2_\sigma(Q_n)$ with domain $D((-A_n)^{\frac{1}{2}})$ such that $D(-A_n) \subset D((-A_n)^{\frac{1}{2}}) \subset L^2_\sigma(Q_n)$ and it enjoys the following properties:

$$D((-A_n)^{\frac{1}{2}}) = H^1_{0,\sigma}(Q_n) \quad \text{and} \quad \langle (-A_n)^{\frac{1}{2}} \mathbf{u}, (-A_n)^{\frac{1}{2}} \mathbf{v}\rangle = \langle \nabla \mathbf{u}, \nabla \mathbf{v}\rangle.$$

We now consider the following abstract Cauchy problem involving the operator $A_n$:

$$(\bar{P}_n) \begin{cases} \frac{\partial \mathbf{u}_n}{\partial t} = A_n \mathbf{u}_n + \mathcal{P}\mathbf{f}_n & \text{in } (0, T), \\ \mathbf{u}_n(0) = \mathbf{u}^0_n. \end{cases}$$

From Theorem 3.8 in [26], the following existence and uniqueness result for the solution of problem $(\bar{P}_n)$ holds.

**Theorem 3.3** *Let $\mathbf{f}_n \in L^2(0,T;L^2(Q_n)^3)$ and $\mathbf{u}_n^0 \in D((-A_n)^{\frac{1}{2}})$. For every $n \in \mathbb{N}$ we set*

$$\mathbf{u}_n(t) = T_n(t)\mathbf{u}_n^0 + \int_0^t T_n(t-s)\mathcal{P}\mathbf{f}_n(s)\,\mathrm{d}s, \tag{3.3}$$

*where $T_n(t)$ is the analytic semigroup generated by $A_n$. Then $\mathbf{u}_n$ is the unique mild solution of $(\bar{P}_n)$, i.e.*

$$\mathbf{u}_n \in H^1(0,T;L^2_\sigma(Q_n)) \bigcap L^2(0,T;D(A_n)),$$

$$\frac{\partial \mathbf{u}_n}{\partial t}(t) = A_n\mathbf{u}_n(t) + \mathcal{P}\mathbf{f}_n(t) \text{ for almost every } t \in [0,T] \text{ and } \mathbf{u}_n(0) = \mathbf{u}_n^0.$$

*Moreover, there exists a constant $C$ independent from $n$ such that the following inequality holds:*

$$\|\mathbf{u}_n\|_{H^1(0,T;L^2_\sigma(Q_n))} + \|\mathbf{u}_n\|_{L^2(0,T;D(A_n))} \leq C\|\mathbf{f}_n\|_{L^2(0,T;L^2(Q_n)^3)} + \|\mathbf{u}_n^0\|_{D((-A_n)^{\frac{1}{2}})}. \tag{3.4}$$

We now prove that the unique mild solution $\mathbf{u}_n$ of the abstract Cauchy problem $(\bar{P}_n)$ is also the unique weak solution of $(\tilde{P}_n)$. We recall the definition of weak solution and of associated pressure given in [32, Chapter IV, Definition 2.1.1].

**Definition 3.4** Let $\mathbf{f}_n \in L^2(0,T;L^2(Q_n)^3)$ and $\mathbf{u}_n^0 \in L^2_\sigma(Q_n)$. A function $\mathbf{u}_n \in L^2(0,T;H^1_{0,\sigma}(Q_n))$ is called a weak solution of the Stokes system $(\tilde{P}_n)$ if and only if

$$-\int_0^T\int_{Q_n} \mathbf{u}_n \cdot \frac{\partial \mathbf{v}}{\partial t}\,\mathrm{d}\mathcal{L}\,\mathrm{d}t + \int_0^T\int_{Q_n} \nabla\mathbf{u}_n \cdot \nabla\mathbf{v}\,\mathrm{d}\mathcal{L}\,\mathrm{d}t = \int_{Q_n} \mathbf{u}_0 \cdot \mathbf{v}(0)\,\mathrm{d}\mathcal{L} + \int_0^T\int_{Q_n} \mathbf{f}_n \cdot \mathbf{v}\,\mathrm{d}\mathcal{L}\,\mathrm{d}t$$

for every $\mathbf{v} \in C_0^\infty([0,T);C^\infty_{0,\sigma}(Q_n))$. A distribution $p_n$ in $[0,T)\times Q_n$ is called an associated pressure of a weak solution $\mathbf{u}_n$ if and only if

$$\frac{\partial \mathbf{u}_n}{\partial t}(t,x) - \Delta\mathbf{u}_n(t,x) + \nabla p_n(t,x) = \mathbf{f}_n(t,x)$$

is satisfied in the sense of distributions.

**Theorem 3.5** *For each $n \in \mathbb{N}$, if $\mathbf{f}_n \in L^2(0,T;L^2(Q_n)^3)$ and $\mathbf{u}_n^0 \in L^2_\sigma(Q_n)$, then the unique mild solution $\mathbf{u}_n$ of problem $(\bar{P}_n)$ given by (3.3) is the unique weak solution of the Stokes system $(\tilde{P}_n)$ in the sense of Definition 3.4. Moreover, there exists a unique pressure $p_n \in L^2(0,T;L^2(Q_n))$ associated with the solution $\mathbf{u}_n$ such that, for a.e. $t \in [0,T]$,*

$$\int_{Q_n} p_n(t,x)\,\mathrm{d}\mathcal{L} = 0.$$

For the proof, see Theorem 3.11 and Section 3.3 in [26].

We now introduce an equivalent weak formulation for $(\tilde{P}_n)$ in view of the numerical approximation of the pre-fractal problems by mixed methods, i.e. FEM in space and finite differences in time. We refer to [30] and [36] for details.

We point out that the space $H^1_{0,\sigma}(Q_n)$ coincides with the space

$$V := \{\mathbf{u} \in H^1_0(Q_n)^3 \, : \, \operatorname{div} \mathbf{u} = 0\}$$

for bounded Lipschitz domains (see Lemma 1.2.1, page 111 in [32]). Moreover, from Lemma 2.5.3, page 82 in [32] we have that $L^2_\sigma(Q_n) = \{\mathbf{u} \in L^2(Q_n)^3 \, : \operatorname{div} \mathbf{u} = 0 \, , \, \nu \cdot \mathbf{u}|_{\partial Q_n} = 0\}$.

Let $\mathbf{u}_n \in V$ be the weak solution of $(\bar{P}_n)$. We multiply the first equation in $(\bar{P}_n)$ by a function $\mathbf{v} \in V$ not depending on time and we integrate on $Q_n$. By using the following property of the Leray–Helmholtz projection (see Lemma 2.5.2, Chapter II in [32])

$$(\mathrm{Id} - \mathcal{P})\mathbf{f} = \nabla p,$$

and integrating by parts, we obtain for a.e. $t \in [0, T)$

$$\int_{Q_n} \frac{\partial \mathbf{u}_n}{\partial t} \cdot \mathbf{v} \, d\mathcal{L} + \int_{Q_n} \nabla \mathbf{u}_n \cdot \nabla \mathbf{v} \, d\mathcal{L} = \int_{Q_n} \mathbf{f}_n \cdot \mathbf{v} \, d\mathcal{L}. \tag{3.5}$$

The existence and uniqueness of the weak solution in $V$ of this problem (which we still denote by $\mathbf{u}_n$) follows from [15, Chapter XIX, Section 2].

This functional setting is not convenient for the numerical approximation of $(\tilde{P}_n)$. Indeed, it is difficult to find a finite dimensional space of divergence-free vector valued functions which allows us to find a good convergence behavior of the approximation error. Moreover, it can be very complicated to construct a basis of this finite dimensional space of divergence-free functions.

We then consider an alternative weak formulation of the Stokes system $(\tilde{P}_n)$. We introduce the space

$$\mathcal{Q} = \left\{ v \in L^2(Q_n) \, : \, \int_{Q_n} v \, d\mathcal{L} = 0 \right\}.$$

By multiplying the first and the second equations of $(\tilde{P}_n)$ by a function $\mathbf{v} \in H^1_0(Q_n)^3$ and integrating on $Q_n$, we get the following weak formulation of $(\tilde{P}_n)$: for a.e. $t \in [0, T)$, given $\mathbf{f}_n \in L^2(Q_n)^3$ and $\mathbf{u}^0_n \in V$, find $\mathbf{u}_n \in H^1_0(Q_n)^3$ and $p_n \in \mathcal{Q}$ such that

$$(P_n) \begin{cases} \dfrac{\partial}{\partial t} \displaystyle\int_{Q_n} \mathbf{u}_n \cdot \mathbf{v} \, d\mathcal{L} + \int_{Q_n} \nabla \mathbf{u}_n \cdot \nabla \mathbf{v} \, d\mathcal{L} - \int_{Q_n} p_n \operatorname{div} \mathbf{v} \, d\mathcal{L} = \int_{Q_n} \mathbf{f}_n \cdot \mathbf{v} \, d\mathcal{L} & \forall \mathbf{v} \in H^1_0(Q_n)^3, \\ \displaystyle\int_{Q_n} q \operatorname{div} \mathbf{u}_n \, d\mathcal{L} = 0 & \forall q \in \mathcal{Q}, \\ \mathbf{u}_n(0, x) = \mathbf{u}^0_n(x) & \text{on } Q_n. \end{cases} \tag{3.6}$$

We point out that the second equation in $(P_n)$ implies that $\operatorname{div} \mathbf{u}_n = 0$ almost everywhere. Problem $(P_n)$ admits a unique weak solution $\mathbf{u}_n \in H_0^1(Q_n)^3$ and an associate pressure $p_n \in \mathcal{Q}$ (see [15, Chapter XIX]). The solution $(\mathbf{u}_n, p_n)$ of (3.6) is also a solution of (3.5) provided that the conditions of [26, Section 3.3] hold. The converse is also true (see [30, Chapters 10 and 13]).

## 4 Regularity in Weighted Sobolev Spaces

We now focus on proving regularity results in weighted Sobolev spaces for the couple $(\mathbf{u}_n, p_n)$, where $\mathbf{u}_n$ is the weak solution of problem $(P_n)$ and $p_n$ is its associated pressure. Such regularity results are needed for the numerical approximation of $(P_n)$; indeed, since the domain $Q_n$ is not convex, we have to construct a suitable triangulation of our domain for obtaining an optimal rate of convergence of the approximation error. The construction of the mesh will rely on this weighted Sobolev regularity. We recall that the interior angles in $Q_n$ have opening equal to $\frac{\pi}{3}$ or $\frac{4}{3}\pi$.

Following the notations of [3], for $\beta \in \mathbb{R}$ and $l \in \mathbb{N}$, we introduce the weighted Sobolev space $V_\beta^l(Q_n)$ of functions for which the following norm is finite:

$$\|u\|^2_{V_\beta^l(Q_n)} := \sum_{|k| \leq l} \int_{Q_n} r^{2(\beta - l + |k|)} |D^k u|^2 \, d\mathcal{L},$$

where $r = r(x)$ is the distance of $x \in Q_n$ from the edges.

It is known in literature (see e.g. [29]) that the regularity of the solution of Stokes problems in polyhedral domains is related to the smallest positive solution $\lambda$ of the following equation:

$$\sin(\lambda\omega) = -\lambda \sin \omega, \tag{4.1}$$

where $\omega$ is the opening of the "worst" interior angle at the edge.

We rephrase Eq. (4.1) in our setting, by recalling that for $Q_n$ $\omega = \frac{4}{3}\pi$. Let $\lambda > 0$ be the (unique) positive solution of the equation

$$\sin\left(\lambda \frac{4\pi}{3}\right) = \lambda \frac{\sqrt{3}}{2}; \tag{4.2}$$

numerically, $\lambda \sim 0.61572$. From Theorem 2.1 in [3] the following result holds.

**Theorem 4.1** *Let $(\mathbf{u}_n, p_n)$ be the solution of $(P_n)$ and its associated pressure. Then for a.e. $t \in [0, T)$ $\mathbf{u}_n \in V^2_\beta(Q_n)^3$, $p_n \in V^1_\beta(Q_n)$ for every $\beta \in (1-\lambda, 1)$, $\partial_{x_3}\mathbf{u}_n \in V^1_0(Q_n)^3$ and $\partial_{x_3} p_n \in L^2(Q_n)$. Moreover, the following a priori estimate holds:*

$$\|\mathbf{u}_n\|_{V^2_\beta(Q_n)^3} + \|\partial_{x_3}\mathbf{u}_n\|_{V^1_0(Q_n)^3} + \|p_n\|_{V^1_\beta(Q_n)} + \|\partial_{x_3} p_n\|_{L^2(Q_n)} \leq C\|\mathbf{f}_n\|_{L^2(Q_n)}, \tag{4.3}$$

*where $C$ depends on $Q_n$ and $\beta$.*

Moreover, the following regularity result in fractional Sobolev spaces holds. For the proof, we refer to Proposition 4.15 in [22].

**Theorem 4.2** *Let $(\mathbf{u}_n, p_n)$ be as in Theorem 4.1. Then $\mathbf{u}_n \in H^{2-\beta}(Q_n)^3$ and $p_n \in H^{1-\beta}(Q_n)$ for every $\beta \in (1-\lambda, 1)$.*

# 5 Mean Shear Stress

Aim of this section is to prove that the average wall shear stress vanishes along the fractal boundary. For the sake of simplicity, we will confine ourselves to the stationary case and we will focus on an isolated system, where the external force contributions vanish. Under the latter assumption the balance force reads

$$\int_{\partial Q_n} \mathbf{t}\left(\hat{\mathbf{n}}\right)\, \mathrm{d}S = 0, \tag{5.1}$$

where $\mathbf{t}$ is the stress vector depending on the unit-length direction vector $\hat{\mathbf{n}}$. Due to the Cauchy's stress theorem, the vector $\mathbf{t}$ can be written in terms of a symmetric second-order tensor field $\mathbf{T}$ as $\mathbf{t}(\hat{\mathbf{n}}) = \mathbf{T} \cdot \hat{\mathbf{n}}$, leading to

$$\int_{\partial Q_n} \mathbf{T} \cdot \hat{\mathbf{n}}\, \mathrm{d}S = 0. \tag{5.2}$$

Since $\partial Q_n = S_n \cup (\Omega_n \times \{0\}) \cup (\Omega_n \times \{1\})$ the above integral factorizes as

$$\int_{S_n} \mathbf{T} \cdot \hat{\mathbf{n}}\, \mathrm{d}\sigma + \int_{\Omega_n \times \{0\}} \mathbf{T} \cdot \hat{\mathbf{n}}\, \mathrm{d}x_1 \mathrm{d}x_2 + \int_{\Omega_n \times \{1\}} \mathbf{T} \cdot \hat{\mathbf{n}}\, \mathrm{d}x_1 \mathrm{d}x_2 = 0. \tag{5.3}$$

For Newtonian incompressible fluids, the stress tensor can be written as $\mathbf{T} = -p\mathbf{1} + \mu(\nabla\mathbf{u} + \nabla\mathbf{u}^T)$, where $\mathbf{1}$ is the identity tensor, $p$ is the hydrodynamic pressure and $\mu$ is the viscosity coefficient. Clearly, in the latter case, the force balance is established between the viscous forces and the pressure field.

A particularly impressive form of the above balance is obtained when considering a Hagen–Poiseuille-like flow, where the cylindrical surface $Q_n$ has as predominant dimension the one represented by its axis. Hence, by denoting as $u_{\perp,n}$ the velocity component aligned with the cylindric axis, and with $(x_1, x_2)$ the coordinates in a generic section perpendicular to the axis, the solution $\mathbf{u}_n$ of the Stokes problem and its associated pressure $p_n$ take the simplified form (see e.g. [34])

$$\mathbf{u}_n = \left(0, 0, u_{\perp,n}\,(x_1, x_2)\right), \qquad p_{\perp,n}(x_3) = p_n(0, 0, x_3).$$

We formally define by

$$\langle \tau_{w,n} \rangle_{K_n} := \frac{1}{|K_n|} \oint_{K_n} \mu \nabla u_{\perp,n} \cdot \hat{\mathbf{n}}\, \mathrm{d}\ell$$

the "average wall shear stress." The following result holds.

**Theorem 5.1** *Let $\tilde{p}_n$ denote the trivial extension of $p_n$ to $Q$ and $\beta \in (1 - \lambda, \frac{1}{2})$. Let us suppose that there exists a constant $M > 0$ independent from $n$ such that*

$$\|\tilde{p}_n(0, 0, x_3)\|_{H^{1-\beta}(Q)} \leq M. \tag{5.4}$$

*Then there exists a positive constant $C$ independent from $n$ such that*

$$|\langle \tau_{w,n} \rangle_{K_n}| \leq C \frac{4}{\sqrt[4]{3}} \left(\frac{3}{4}\right)^n |\Omega|, \tag{5.5}$$

*hence the average wall shear stress $\langle \tau_{w,n} \rangle_{K_n}$ vanishes as $n \to +\infty$.*

***Proof*** Under the above assumptions, the balance in (5.3) takes the form

$$\int_I \mathrm{d}x_3 \oint_{K_n} \mu \nabla u_{\perp,n} \cdot \hat{\mathbf{n}}\, \mathrm{d}\ell + \int_{\Omega_n \times \{0\}} p_{\perp,n}(x_3)\, \mathrm{d}x_1 \mathrm{d}x_2 - \int_{\Omega_n \times \{1\}} p_{\perp,n}(x_3)\, \mathrm{d}x_1 \mathrm{d}x_2 = 0.$$

From Theorem 4.1 all the above integrals are well defined for the solution $\mathbf{u}_n$ and its associated pressure $p_n$. By denoting with $\tau_{w,n} = \mu \nabla u_{\perp,n} \cdot \hat{\mathbf{n}}$ the wall shear stress, since $I = [0, 1]$, one has

$$\oint_{K_n} \tau_{w,n}\, \mathrm{d}\ell + |\Omega_n| \left(p_{\perp,n}|_{\{x_3=0\}} - p_{\perp,n}|_{\{x_3=1\}}\right) = 0,$$

where $|\mathcal{A}|$ denotes the Lebesgue measure of a subset $\mathcal{A} \subset \mathbb{R}^N$.

By dividing both members of the above equation by $|K_n|$, we obtain

$$\langle \tau_{w,n} \rangle_{K_n} = \frac{1}{|K_n|} \oint_{K_n} \tau_{w,n}\, \mathrm{d}\ell = -\frac{|\Omega_n|}{|K_n|} \left(p_{\perp,n}|_{\{x_3=0\}} - p_{\perp,n}|_{\{x_3=1\}}\right) \tag{5.6}$$

$$= \left(\frac{3}{4}\right)^n |\Omega_n| \left(p_{\perp,n}|_{\{x_3=1\}} - p_{\perp,n}|_{\{x_3=0\}}\right). \tag{5.7}$$

Since $|\Omega_n| \le |\Omega|$, from (5.7) we obtain

$$|\langle \tau_{w,n} \rangle_{K_n}| \le \left(\frac{3}{4}\right)^n |\Omega| \left( |p_{\perp,n}|_{\{x_3=1\}}| + |p_{\perp,n}|_{\{x_3=0\}}| \right). \tag{5.8}$$

We recall that from Theorem 4.2 $p_n \in H^{1-\beta}(Q_n)$ for every $\beta \in (1-\lambda, 1)$; hence $\tilde{p}_n \in H^{1-\beta}(Q)$ for every $\beta \in (1-\lambda, 1)$.
From Sobolev embeddings and the trace Theorem 2.1, we obtain for every $\beta \in (1-\lambda, \frac{1}{2})$ that

$$\begin{aligned} p_{\perp,n}|_{\{x_3=0\}} &= \frac{1}{|\Omega_n|} \int_{\Omega_n \times \{0\}} p_{\perp,n}(x_3)\, \mathrm{d}x_1 \mathrm{d}x_2 \le \frac{1}{|\Omega_n|^{\frac{1}{2}}} \|p_{\perp,n}(x_3)\|_{L^2(\Omega_n \times \{0\})} \\ &= \frac{1}{|\Omega_n|^{\frac{1}{2}}} \|p_n(0,0,x_3)\|_{L^2(\Omega_n \times \{0\})} \\ &= \frac{1}{|\Omega_n|^{\frac{1}{2}}} \|\tilde{p}_n(0,0,x_3)\|_{L^2(\Omega \times \{0\})} \le \frac{C_{\mathrm{Sob}}}{|\Omega_n|^{\frac{1}{2}}} \|\tilde{p}_n(0,0,x_3)\|_{H^{\frac{1}{2}-\beta}(\Omega \times \{0\})} \\ &\le \frac{C_{\mathrm{Sob}} C_{\mathrm{Tr}}}{|\Omega_n|^{\frac{1}{2}}} \|\tilde{p}_n(0,0,x_3)\|_{H^{1-\beta}(Q)}. \end{aligned} \tag{5.9}$$

Since the area of $\Omega_n$ is bigger than the area of the equilateral triangle of vertices $A_1$, $A_3$ and $A_5$ (see Sect. 2), which in particular is $\frac{\sqrt{3}}{4}$, from (5.4), we have that

$$p_{\perp,n}|_{\{x_3=0\}} \le C \frac{2}{\sqrt[4]{3}}, \tag{5.10}$$

where the constant $C$ depends on $Q$ and $M$ but it is independent from $n$.
We can estimate $p_{\perp,n}|_{\{x_3=1\}}$ in the same way. Hence, from (5.8) we get

$$|\langle \tau_{w,n} \rangle_{K_n}| \le C \frac{4}{\sqrt[4]{3}} \left(\frac{3}{4}\right)^n |\Omega|. \tag{5.11}$$

Therefore $\langle \tau_{w,n} \rangle_{K_n} \to 0$ as $n \to +\infty$, thus proving that the average wall shear stress decreases if we increase the approximation of the pre-fractal $n$. □

*Remark 5.2* In the proof of Theorem 4.9 in [26], it is shown that $\tilde{p}_n$ is equibounded in $L^2(Q)$. Hence, it seems reasonable to suppose in Theorem 5.1 that $\|\tilde{p}_n(0,0,x_3)\|_{H^{1-\beta}(Q)}$ is equibounded; indeed, in our case $\tilde{p}_n(0,0,x_3)$ depends only on the $x_3$ variable, which takes values in $I = [0,1]$.

*Remark 5.3* It is worth noticing that the hypotheses on the structure of the velocity and pressure fields (which are physically reasonable in laminar pipe flows) are crucial in order to enforce the balance between the pressure field and the wall shear stress. Only in the latter case the stress on $\Omega_n \times \{0\}$ and $\Omega_n \times \{1\}$ is fixed by the pressure value, and viscous contributions vanish. However, laminar pipe flows are present in several interesting scenery, e.g. blood flow in capillaries and microfluidic devices, making the theoretical results of potential interest for many practical applications.

## 6 Numerical Approximation

We now consider the numerical approximation of problem $(P_n)$ by FEM in space and FD in time. We construct an ad hoc mesh which takes into account the singularities of $\mathbf{u}_n$ in the vertices and in the edges of $\partial Q_n$, in order to obtain an optimal a priori error estimate.

In [4], the case of two-dimensional pre-fractal domains was studied. By adapting the results of [4] to the present case, we first construct a mesh of size $h$ for the two-dimensional domains $\Omega_n$, and then following [3], we extend it to the 3D case. We point out that any mesh generated by applying the proposed algorithm is compliant with the Grisvard conditions [17].

Let $\{T\}$ be a regular triangulation of $\Omega_n$, $\tilde{\mu} \in [0, 1)$ be the grading parameter, $r_T$ be the distance of the triangle $T$ from the re-entrant corner of opening $\frac{4\pi}{3}$, i.e.

$$r_T = \inf_{(x_1,x_2)\in T} \sqrt{x_1^2 + x_2^2},$$

and $h_T$ be the diameter of $T$. We assume that

$$h_T \le \begin{cases} \tilde{\lambda} h^{\frac{1}{1-\tilde{\mu}}} & \text{for } r_T = 0, \\ \tilde{\lambda} h r_T^{\tilde{\mu}} & \text{for } 0 < r_T \le R, \\ \tilde{\lambda} h & \text{for } r_T > R, \end{cases}$$

for some constant $R > 0$, where $\tilde{\lambda}$ is the regularity constant of the mesh.

We now extend this two-dimensional mesh in the third dimension by using a uniform mesh size $h$. Let $\{E\}$ be an element of the three-dimensional mesh obtained in such way; we denote by $r_E$ the distance of $E$ from the edges and by $h_{1,E}$ the length of the projection of $E$ on the $x_1$ axis (analogously, we define $h_{2,E}$ and $h_{3,E}$). Then the element sizes satisfy the following:

$$h_{1,E}, h_{2,E} \le \begin{cases} \tilde{\lambda} h^{\frac{1}{1-\tilde{\mu}}} & \text{for } r_E = 0, \\ \tilde{\lambda} h r_E^{\tilde{\mu}} & \text{for } 0 < r_E \le R, \\ \tilde{\lambda} h & \text{for } r_E > R, \end{cases} \qquad h_{3,E} \le \tilde{\lambda} h.$$

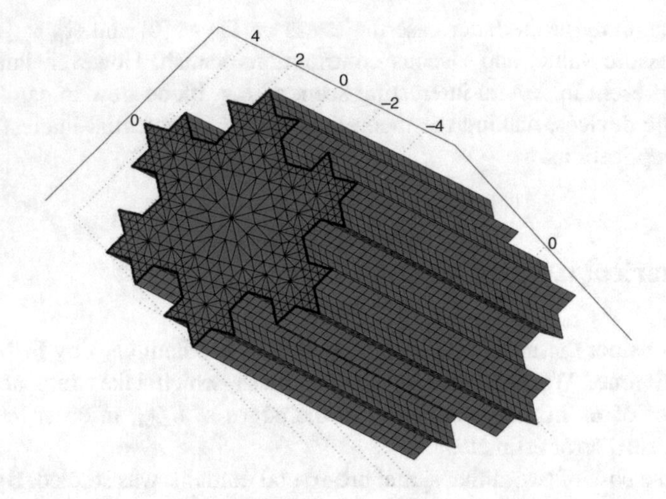


**Fig. 2** The mesh of $Q_n$, for $n = 2$

We denote by $T_{n,h} = \{E\}$ the three-dimensional mesh obtained in this way and by $F$ the faces of the elements. We point out that following this construction we obtain either a pentahedral or a tetrahedral triangulation. In view of the numerical simulations, we choose the pentahedral one, but the results of this section hold also for tetrahedral meshes. The mesh obtained by using this procedure is shown in Fig. 2.

We now discretize problem $(P_n)$ in space first. Let $\mathbb{P}_1$ denote the space of polynomial functions of degree one. We denote by $X_{n,h}$ the finite element space

$$X_{n,h} = \left\{ \mathbf{v} \in L^2(Q_n)^3 \;:\; \mathbf{v}|_E \in (\mathbb{P}_1 \oplus \text{span}\{x_3^2\})^3 \,\forall\, E \in T_{n,h} \text{ and } \int_F [\mathbf{v}] = 0 \,\forall\, F \right\},$$

where $[\mathbf{v}]$ denotes the jump of $\mathbf{v}$ on the faces $F$; for boundary faces $[\mathbf{v}]$ is identified with $\mathbf{v}$.

We denote by $M_{n,h}$ the subspace of piecewise constant functions

$$M_{n,h} = \left\{ q \in L^2(Q_n) \;:\; q|_E \in \mathbb{P}_0 \,\forall\, E \in T_{n,h} \text{ and } \int_{Q_n} q = 0 \right\}.$$

We point out that $X_{n,h} \not\subseteq H_0^1(Q_n)^3$. Moreover, we equip $X_{n,h}$ with the following norm:

$$\|\mathbf{u}\|^2_{X_{n,h}} := \sum_E \int_E \nabla\mathbf{u} \cdot \nabla\mathbf{u}\, \mathrm{d}\mathcal{L}.$$

We denote by $I_{n,h}$ the Crouzeix–Raviart interpolant polynomial (see [14]) $I_{n,h}\colon H_0^1(Q_n)^3 \to X_{n,h}$ defined elementwise by

$$\int_F \mathbf{u}\,\mathrm{d}\mathcal{S} = \int_F I_{n,h}\mathbf{u}\,\mathrm{d}\mathcal{S} \qquad \forall\, F \subset \partial E,\ \forall\, E \in T_{n,h}.$$

In [2] it is proved that $I_{n,h}$ is well defined. Moreover, it is stable in $H_0^1(Q_n)^3$, i.e.

$$\int_E \nabla(I_{n,h}\mathbf{u})\,\mathrm{d}\mathcal{L} \le C \int_E \nabla\mathbf{u}\,\mathrm{d}\mathcal{L} \quad \forall\, E \in T_{n,h}.$$

If $\mathbf{u}_n = (u_{1,n}, u_{2,n}, u_{3,n})$ is the solution of $(P_n)$, from (4.3) and Theorem 5.1 in [2] we obtain that

$$\|u_{i,n} - I_{n,h}u_{i,n}\|_{X_{n,h}} \le Ch\|\mathbf{f}_n\|_{L^2(Q_n)^3}.$$

We now introduce the following mean value operator:

$$\mathcal{M}_{h,E}\, p = \frac{1}{\mathrm{meas}(E)} \int_E p\,\mathrm{d}\mathcal{L}.$$

We set $\mathcal{M}_h p(x) := \mathcal{M}_{h,E}\, p(x)$ if $x \in E$. By proceeding as in [3], if $p_n$ is the pressure associated with the solution $\mathbf{u}_n$ of $(P_n)$, from (4.3) we have that

$$\|p_n - \mathcal{M}_h\, p_n\|_{L^2(Q_n)} \le Ch\|\mathbf{f}_n\|_{L^2(Q_n)^3}.$$

The semi-discrete approximation problem reads as follows: given $\mathbf{u}^0_{n,h} \in X_{n,h}$, for a.e. $t \in [0, T)$ find $\mathbf{u}_{n,h} \in X_{n,h}$ and $p_{n,h} \in M_{n,h}$ such that

$$(P_{n,h}) \begin{cases} \dfrac{\partial}{\partial t}(\mathbf{u}_{n,h}, \mathbf{v}_h)_{L^2(Q_n)} + a_n(\mathbf{u}_{n,h}, \mathbf{v}_{n,h}) - (p_{n,h}, \operatorname{div}\mathbf{v}_h)_{L^2(Q_n)} & \\ \quad = (\mathbf{f}_n, \mathbf{v}_h)_{L^2(Q_n)} & \forall\, \mathbf{v}_h \in X_{n,h}, \\ (q_h, \operatorname{div}\mathbf{u}_{n,h})_{L^2(Q_n)} = 0 & \forall\, q_h \in M_{n,h}, \\ \mathbf{u}_{n,h}(0) = \mathbf{u}^0_{n,h}. & \end{cases} \tag{6.1}$$

Problem $(P_{n,h})$ is a system of ordinary differential equations. It admits a unique weak solution given by the couple $(\mathbf{u}_{n,h}, p_{n,h})$, see e.g. [15].

From Lemma 3.1 in [3], we have that the following *inf-sup condition* holds: for every $q_h \in M_{n,h}$ there exists a function $\mathbf{v}_h \in X_{n,h}$, with $\mathbf{v}_h \neq 0$, such that

$$(q_h, \operatorname{div}\mathbf{v}_h)_{L^2(Q_n)} \ge C\|q_h\|_{L^2(Q_n)}\|\mathbf{v}_h\|_{X_{n,h}}, \tag{6.2}$$

where $C$ is a constant depending on $Q_n$ and independent from $h$.

Using the above results, we obtain the following optimal a priori error estimate (see Theorem 3.1 in [3]).

**Theorem 6.1** *Let* $(\mathbf{u}_n, p_n)$ *be the solution of problem* $(P_n)$ *and its associated pressure, respectively. Let* $(\mathbf{u}_{n,h}, p_{n,h})$ *be the semi-discrete solution of problem* $(P_{n,h})$*. If the grading parameter of the mesh* $\tilde{\mu}$ *is such that* $\tilde{\mu} < \lambda$*, where* $\lambda$ *is given by* (4.2)*, then for a.e.* $t \in [0, T)$ *the following estimate holds with a constant* $C_1$ *independent from* $h$*:*

$$\|\mathbf{u}_n - \mathbf{u}_{n,h}\|_{X_{n,h}} + \|p_n - p_{n,h}\|_{L^2(Q_n)} \leq \|\mathbf{u}_n^0 - \mathbf{u}_{n,h}^0\|_{X_{n,h}} + C_1 h \|\mathbf{f}_n\|_{L^2(Q_n)^3}. \tag{6.3}$$

We now approximate problem $(P_{n,h})$ in time by using the implicit Euler method. From now on we suppose that $\frac{\partial \mathbf{f}_n}{\partial t} \in L^2(Q_n)^3$.

Let $\Delta t = \frac{T}{N}$ be the time step size, for some integer $N$, and let $t_l = l\Delta t$, for $l = 0, 1, \ldots, N$. We set

$$V_{n,h} := \left\{ \mathbf{v}_h \in X_{n,h} : \int_{Q_n} q_h \operatorname{div} \mathbf{v}_h \, d\mathcal{L} = 0 \quad \forall q_h \in M_{n,h} \right\}.$$

Let $\mathcal{P}_{n,h} \colon L^2(Q_n)^3 \to V_{n,h}$ be the orthogonal projection. We define $A_{n,h} := -\mathcal{P}_{n,h}\Delta$ the discrete version of the Stokes operator. Since the restriction of $A_{n,h}$ to $V_{n,h}$ is invertible, and its inverse $A_{n,h}^{-1}$ is positive and self-adjoint, we can define the following discrete norms for $r \in \mathbb{R}$ and $\mathbf{v}_h \in V_{n,h}$:

$$\|\mathbf{v}_h\|_r := \|A_{n,h}^{\frac{r}{2}} \mathbf{v}_h\|_{L^2(Q_n)^3}.$$

We point out that the following hold:

$$\|\mathbf{v}_h\|_0 = \|\mathbf{v}_h\|_{L^2(Q_n)^3}, \quad \|\mathbf{v}_h\|_1 = \|\mathbf{v}_h\|_{X_{n,h}}, \quad \|\mathbf{v}_h\|_2 = \|A_{n,h}\mathbf{v}_h\|_{L^2(Q_n)^3}.$$

The fully discretized problem reads as follows: setting $\mathbf{u}_{n,h}^{l=0} = \mathbf{u}_{n,h}(0)$, find $\mathbf{u}_{n,h}^l \in X_{n,h}$ and $p_{n,h}^l \in M_{n,h}$ such that

$$(P_{n,h}^l) \begin{cases} \dfrac{(\mathbf{u}_{n,h}^l - \mathbf{u}_{n,h}^{l-1}, \mathbf{v}_h)_{L^2(Q_n)}}{\Delta t} + a_n(\mathbf{u}_{n,h}^l, \mathbf{v}_h) - (p_{n,h}^l, \operatorname{div} \mathbf{v}_h)_{L^2(Q_n)} & \\ \quad = (\mathbf{f}_n(t_l), \mathbf{v}_h)_{L^2(Q_n)} & \forall \mathbf{v}_h \in X_{n,h}, \\ (q_h, \operatorname{div} \mathbf{u}_{n,h}^l)_{L^2(Q_n)} = 0 & \forall q_h \in M_{n,h}. \end{cases} \tag{6.4}$$

We adapt the results of [18] to our case; for the case of Crank–Nicolson, we refer to [19]. From now on, we set

$$d_t\mathbf{w}^l_{n,h} = \frac{\mathbf{w}^l_{n,h} - \mathbf{w}^{l-1}_{n,h}}{\Delta t}.$$

**Theorem 6.2 (Stability)** *The fully discretized solution* $\mathbf{u}^l_{n,h}$ *of* $(P^l_{n,h})$ *satisfies the following estimate for every* $m = 0, \dots, N$*:*

$$\|\mathbf{u}^m_{n,h}\|^2_{X_{n,h}} + \Delta t \sum_{l=1}^{m} \|\mathbf{u}^l_{n,h}\|^2_2 \leq \|\mathbf{u}^0_{n,h}\|^2_{X_{n,h}} + C \sup_{t\in[0,T]} \|\mathbf{f}_n(t)\|^2_{L^2(Q_n)^3}, \tag{6.5}$$

*where C depends on T and it is independent from* $\Delta t$, $N$ *and* $h$.

***Proof*** We take $v_h = 2\Delta t A_{n,h}\mathbf{u}^l_{n,h}$ as test function in the first equation of $(P^l_{n,h})$. By using vectorial identities and the properties of $A_{n,h}$, we have

$$\|\mathbf{u}^l_{n,h}\|^2_1 - \|\mathbf{u}^{l-1}_{n,h}\|^2_1 + \Delta t^2\|d_t\mathbf{u}^l_{n,h}\|^2_1 + 2\Delta t\|A_{n,h}\mathbf{u}^l_{n,h}\|^2_{L^2(Q_n)^3} = 2\Delta t(\mathbf{f}_n(t_l), A_{n,h}\mathbf{u}^l_{n,h})_{L^2(Q_n)}. \tag{6.6}$$

From Cauchy–Schwarz and Young inequalities it follows that

$$\|\mathbf{u}^l_{n,h}\|^2_1 - \|\mathbf{u}^{l-1}_{n,h}\|^2_1 + \Delta t\|A_{n,h}\mathbf{u}^l_{n,h}\|^2_{L^2(Q_n)^3} \leq \Delta t\|\mathbf{f}_n(t_l)\|^2_{L^2(Q_n)^3}. \tag{6.7}$$

By summing on $l$ from 1 to $m$ and taking the supremum for $t \in [0, T]$ on the right-hand side we get the thesis. □

We set $\sigma(t) = \min\{1, t\}$, $\mathbf{e}^l_{n,h} := \mathbf{u}^l_{n,h} - \mathbf{u}_{n,h}(t_l)$ and $\eta^l_{n,h} := p^l_{n,h} - p_{n,h}(t_l)$ for every $l = 1, \dots, N$. By calculating $(P_{n,h})$ in $t = t_l$ and by subtracting it to $(P^l_{n,h})$ we get

$$\begin{cases} (d_t\mathbf{e}^l_{n,h}, \mathbf{v}_h)_{L^2(Q_n)} + a_n(\mathbf{e}^l_{n,h}, \mathbf{v}_{n,h}) - (\eta^l_{n,h}, \operatorname{div}\mathbf{v}_h)_{L^2(Q_n)} = (E^l_{n,h}, \mathbf{v}_h)_{L^2(Q_n)} & \forall\, \mathbf{v}_h \in X_{n,h}, \\ (q_h, \operatorname{div}\mathbf{e}^l_{n,h})_{L^2(Q_n)} = 0 & \forall\, q_h \in M_{n,h}, \end{cases} \tag{6.8}$$

where the "residual term" $E^l_{n,h}$ is defined as follows:

$$(E^l_{n,h}, \mathbf{v}_h)_{L^2(Q_n)} := \frac{1}{\Delta t}\int_{t_{l-1}}^{t_l} (t - t_{l-1}) \left(\frac{\partial^2\mathbf{u}_{n,h}(t)}{\partial t^2}, \mathbf{v}_h\right)_{L^2(Q_n)} dt. \tag{6.9}$$

We recall an a priori estimate for the semi-discrete solution $\mathbf{u}_{n,h}$ of $(P_{n,h})$. For the proof, we refer to Proposition 3.2 in [19].

**Proposition 6.3** *Let $\mathbf{u}_{n,h}$ be the solution of $(P_{n,h})$. Let us suppose that*

$$\left\| \frac{\partial \mathbf{f}_n}{\partial t} \right\|^2_{L^2(Q_n)^3} \leq \tilde{C},$$

*where $\tilde{C}$ is independent from n and h. Then*

$$\sum_{r=0}^{2} \sigma^r(t) \left\| \frac{\partial \mathbf{u}_{n,h}}{\partial t} \right\|^2_r \leq C \left( \|\mathbf{u}_{n,h}(0)\|^2_{X_{n,h}} + \int_0^t \left( \|\mathbf{f}_n\|^2_{L^2(Q_n)^3} + \left\| \frac{\partial \mathbf{f}_n}{\partial t} \right\|^2_{L^2(Q_n)^3} \right) \mathrm{d}s \right).$$

We prove some preliminary lemmas. For the sake of completeness, we give sketches of the proofs.

**Lemma 6.4** *Let us suppose that*

$$\left\| \frac{\partial \mathbf{f}_n}{\partial t} \right\|^2_{L^2(Q_n)^3} + \left\| \frac{\partial^2 \mathbf{f}_n}{\partial t^2} \right\|^2_{L^2(Q_n)^3} \leq \tilde{C},$$

*where $\tilde{C}$ is independent from n and h. Under the assumptions of Theorem 6.2, the residual term $E^l_{n,h}$ satisfies the following estimates:*

$$\Delta t \sum_{l=1}^{m} \|A^{-1}_{n,h} \mathcal{P}_{n,h} E^l_{n,h}\|^2_{L^2(Q_n)^3} \leq C \Delta t^2 \left( \|\mathbf{u}_{n,h}(0)\|^2_{X_{n,h}} + \int_0^T \left( \|\mathbf{f}_n\|^2_{L^2(Q_n)^3} + \left\| \frac{\partial \mathbf{f}_n}{\partial t} \right\|^2_{L^2(Q_n)^3} \right) \mathrm{d}s \right), \tag{6.10}$$

$$\Delta t \sum_{l=1}^{m} \|A^{-\frac{1}{2}}_{n,h} \mathcal{P}_{n,h} E^l_{n,h}\|^2_{L^2(Q_n)^3} \leq C \Delta t \left( \|\mathbf{u}_{n,h}(0)\|^2_{X_{n,h}} + \int_0^T \left( \|\mathbf{f}_n\|^2_{L^2(Q_n)^3} + \left\| \frac{\partial \mathbf{f}_n}{\partial t} \right\|^2_{L^2(Q_n)^3} \right) \mathrm{d}s \right), \tag{6.11}$$

$$\Delta t \sum_{l=1}^{m} \sigma(t_l) \|A^{-\frac{1}{2}}_{n,h} \mathcal{P}_{n,h} E^l_{n,h}\|^2_{L^2(Q_n)^3} \leq C \Delta t^2 \left( \|\mathbf{u}_{n,h}(0)\|^2_{X_{n,h}} + \int_0^T \left( \|\mathbf{f}_n\|^2_{L^2(Q_n)^3} + \left\| \frac{\partial \mathbf{f}_n}{\partial t} \right\|^2_{L^2(Q_n)^3} \right) \mathrm{d}s \right), \tag{6.12}$$

$$\sigma^3(t_m) \|E^m_{n,h}\|^2_{L^2(Q_n)^3} + \Delta t \sum_{l=2}^{m} \sigma^2(t_l) \|E^l_{n,h}\|^2_{L^2(Q_n)^3} \leq C \Delta t^2 \|\mathbf{u}_{n,h}(0)\|^2_{X_{n,h}}$$

$$+ C \Delta t^2 \int_0^T \left( \|\mathbf{f}_n\|^2_{L^2(Q_n)^3} + \left\| \frac{\partial \mathbf{f}_n}{\partial t} \right\|^2_{L^2(Q_n)^3} \right) \mathrm{d}s, \quad m = 2, \ldots, N, \tag{6.13}$$

$$\Delta t \sum_{l=3}^{m} \sigma^3(t_l) \|A^{-\frac{1}{2}}_{n,h} \mathcal{P}_{n,h} d_t E^l_{n,h}\|^2_{L^2(Q_n)^3} \leq C \Delta t^2 \|\mathbf{u}_{n,h}(0)\|^2_{X_{n,h}}$$

$$+ C \Delta t^2 \int_0^T \left( \|\mathbf{f}_n\|^2_{L^2(Q_n)^3} + \left\| \frac{\partial \mathbf{f}_n}{\partial t} \right\|^2_{L^2(Q_n)^3} + \left\| \frac{\partial^2 \mathbf{f}_n}{\partial t^2} \right\|^2_{L^2(Q_n)^3} \right) \mathrm{d}s, \quad m = 3, \ldots, N. \tag{6.14}$$

For the proof we refer to [18, Lemma 6.1] with small suitable changes.

**Lemma 6.5** *Under the assumptions of Lemma 6.4, the following inequality holds for every $m = 1, \dots, N$:*

$$\|\mathbf{e}^m_{n,h}\|^2_{L^2(Q_n)^3} + \Delta t \sum_{l=1}^{m} \left( \Delta t \|d_t \mathbf{e}^l_{n,h}\|^2_{L^2(Q_n)^3} + \|\mathbf{e}^l_{n,h}\|^2_1 \right) \leq C \Delta t \|\mathbf{u}_{n,h}(0)\|^2_{X_{n,h}}$$

$$+ C \Delta t \int_0^T \left( \|\mathbf{f}_n\|^2_{L^2(Q_n)^3} + \left\| \frac{\partial \mathbf{f}_n}{\partial t} \right\|^2_{L^2(Q_n)^3} \right) \mathrm{d}s. \tag{6.15}$$

***Proof*** By taking $\mathbf{v}_h = 2\mathbf{e}^l_{n,h} \Delta t$ as test function in (6.8), noting that this function belongs to $V_{n,h}$ and using Cauchy–Schwarz and Young inequalities, we get

$$\|\mathbf{e}^l_{n,h}\|^2_{L^2(Q_n)^3} - \|\mathbf{e}^{l-1}_{n,h}\|^2_{L^2(Q_n)^3} + \Delta t^2 \|d_t \mathbf{e}^l_{n,h}\|^2_{L^2(Q_n)^3} + c \Delta t \|\mathbf{e}^m_{n,h}\|^2_1 \leq C \Delta t \|A^{-\frac{1}{2}}_{n,h} \mathcal{P}_{n,h} E^l_{n,h}\|^2_{L^2(Q_n)^3}. \tag{6.16}$$

By summing from 1 to $N$ and using (6.11), we get the thesis. □

**Lemma 6.6** *Under the assumptions of Lemma 6.4, the following inequality holds for every $m = 1, \dots, N$:*

$$\sigma(t_m) \|\mathbf{e}^m_{n,h}\|^2_{L^2(Q_n)^3} + c \Delta t \sum_{l=1}^{m} \sigma(t_l) \|\mathbf{e}^l_{n,h}\|^2_1 \leq C \Delta t^2 \|\mathbf{u}_{n,h}(0)\|^2_{X_{n,h}}$$

$$+ C \Delta t^2 \int_0^T \left( \|\mathbf{f}_n\|^2_{L^2(Q_n)^3} + \left\| \frac{\partial \mathbf{f}_n}{\partial t} \right\|^2_{L^2(Q_n)^3} \right) \mathrm{d}s. \tag{6.17}$$

***Proof*** By multiplying (6.16) by $\sigma(t_l)$ and summing on $l$ from 1 to $m$ we get

$$\sigma(t_m) \|\mathbf{e}^m_{n,h}\|^2_{L^2(Q_n)^3} + c \Delta t \sum_{l=1}^{m} \sigma(t_l) \|\mathbf{e}^l_{n,h}\|^2_1 \leq C \Delta t \sum_{l=1}^{m} \sigma(t_l) \|A^{-\frac{1}{2}}_{n,h} \mathcal{P}_{n,h} E^l_{n,h}\|^2_{L^2(Q_n)^3}.$$

From (6.12) we get the thesis.

□

**Lemma 6.7** *Under the assumptions of Lemma 6.4, the following inequality holds for every $m = 1, \dots, N$:*

$$\sigma^2(t_m) \|\mathbf{e}^m_{n,h}\|^2_1 + c \Delta t \sum_{l=2}^{m} \sigma^2(t_l) \|A_{n,h} \mathbf{e}^l_{n,h}\|^2_{L^2(Q_n)^3} \leq C \Delta t^2 \|\mathbf{u}_{n,h}(0)\|^2_{X_{n,h}}$$

$$+ C \Delta t^2 \int_0^T \left( \|\mathbf{f}_n\|^2_{L^2(Q_n)^3} + \left\| \frac{\partial \mathbf{f}_n}{\partial t} \right\|^2_{L^2(Q_n)^3} \right) \mathrm{d}s. \tag{6.18}$$

***Proof*** By taking $\mathbf{v}_h = 2A_{n,h}\mathbf{e}^l_{n,h}\Delta t$ as test function in (6.8), noting that this function belongs to $V_{n,h}$ and using Cauchy–Schwarz and Young inequalities, we get

$$\|\mathbf{e}^l_{n,h}\|_1^2 - \|\mathbf{e}^{l-1}_{n,h}\|_1^2 + \Delta t^2\|d_t\mathbf{e}^l_{n,h}\|_1^2 + c\Delta t\|A_{n,h}\mathbf{e}^l_{n,h}\|^2_{L^2(Q_n)^3} \le C\Delta t\|E^l_{n,h}\|^2_{L^2(Q_n)^3}.$$

By multiplying the above inequality by $\sigma^2(t_l)$, summing on $l$ from 2 to $m$ and using (6.13) we get the thesis. □

By combining (6.17) and (6.18) we obtain the following error estimates for the velocities: for $t_l \in (0, T)$

$$\begin{aligned}&\sigma(t_l)\|\mathbf{e}^l_{n,h}\|^2_{L^2(Q_n)^3} + \sigma^2(t_l)\|\mathbf{e}^l_{n,h}\|^2_{X_{n,h}}\\ &\quad \le C_2\Delta t^2\left(\|\mathbf{u}_{n,h}(0)\|^2_{X_{n,h}} + \int_0^T\left(\|\mathbf{f}_n\|^2_{L^2(Q_n)^3} + \left\|\frac{\partial \mathbf{f}_n}{\partial t}\right\|^2_{L^2(Q_n)^3}\right)\mathrm{d}s\right).\end{aligned} \tag{6.19}$$

where $C_2$ is independent from $t$ and $\Delta t$.

We now focus on obtaining the error estimate for the pressure term $\eta^l_{n,h}$.

**Theorem 6.8** *Under the assumptions of Lemma 6.4, for $l = 1, \dots, N$ we have that*

$$\begin{aligned}&\sigma^3(t_l)\|\eta^l_{n,h}\|^2_{L^2(Q_n)}\\ &\quad \le C\Delta t^2\left(\|\mathbf{u}_{n,h}(0)\|^2_{X_{n,h}} + \int_0^T\left(\|\mathbf{f}_n\|^2_{L^2(Q_n)^3} + \left\|\frac{\partial \mathbf{f}_n}{\partial t}\right\|^2_{L^2(Q_n)^3} + \left\|\frac{\partial^2 \mathbf{f}_n}{\partial t^2}\right\|^2_{L^2(Q_n)^3}\right)\mathrm{d}s\right).\end{aligned} \tag{6.20}$$

***Proof*** Taking $\mathbf{v}_h \in V_{n,h}$, from (6.8) it follows that

$$(d_{tt}\mathbf{e}^l_{n,h}, \mathbf{v}_h)_{L^2(Q_n)} + a_n(d_t\mathbf{e}^l_{n,h}, \mathbf{v}_{n,h}) = (d_t E^l_{n,h}, \mathbf{v}_h)_{L^2(Q_n)}.$$

Setting $\mathbf{v}_h = 2\Delta t d_t\mathbf{e}^l_{n,h}$ in the above equation, it follows that

$$\|d_t\mathbf{e}^l_{n,h}\|^2_{L^2(Q_n)^3} - \|d_t\mathbf{e}^{l-1}_{n,h}\|^2_{L^2(Q_n)^3} + c\Delta t\|d_t\mathbf{e}^l_{n,h}\|_1^2 \le C\Delta t\|A_{n,h}^{-\frac{1}{2}}\mathcal{P}_{n,h}d_t E^l_{n,h}\|^2_{L^2(Q_n)^3}.$$

Multiplying the above inequality by $\sigma^3(t_l)$, summing on $l$ from 3 to $m$ and using (6.14) we get

$$\begin{aligned}&\sigma^3(t_m)\|d_t\mathbf{e}^m_{n,h}\|^2_{L^2(Q_n)^3}\\ &\quad \le C\Delta t^2\left(\|\mathbf{u}_{n,h}(0)\|^2_{X_{n,h}} + \int_0^T\left(\|\mathbf{f}_n\|^2_{L^2(Q_n)^3} + \left\|\frac{\partial \mathbf{f}_n}{\partial t}\right\|^2_{L^2(Q_n)^3} + \left\|\frac{\partial^2 \mathbf{f}_n}{\partial t^2}\right\|^2_{L^2(Q_n)^3}\right)\mathrm{d}s\right).\end{aligned} \tag{6.21}$$

We now estimate the residual term $E^1_{n,h}$. From direct calculations it follows that

$$\|E^1_{n,h}\|_{L^2(Q_n)^3} \leq \left\|\frac{\partial \mathbf{u}_{n,h}(t_1)}{\partial t}\right\|_{L^2(Q_n)^3} + \frac{1}{\sqrt{\Delta t}}\left(\int_{t_0}^{t_1}\left\|\frac{\partial \mathbf{u}_{n,h}(t)}{\partial t}\right\|^2_{L^2(Q_n)^3} \mathrm{d}t\right)^{\frac{1}{2}}.$$

Multiplying the above inequality by $\sigma^3(t_1)$, since $\sigma(t_1) \leq \Delta t$, we get

$$\sigma^3(t_1)\|E^1_{n,h}\|^2_{L^2(Q_n)^3} \leq \sigma^3(t_1)\left\|\frac{\partial \mathbf{u}_{n,h}(t_1)}{\partial t}\right\|^2_{L^2(Q_n)^3} + \sigma^2(t_1)\int_{t_0}^{t_1}\left\|\frac{\partial \mathbf{u}_{n,h}(t)}{\partial t}\right\|^2_{L^2(Q_n)^3} \mathrm{d}t.$$

Since $\sigma^2(t_1) \leq \Delta t^2$, from Proposition 6.3 we obtain

$$\sigma^3(t_1)\|E^1_{n,h}\|^2_{L^2(Q_n)^3} \leq C\Delta t^2\left(\|\mathbf{u}_{n,h}(0)\|^2_{X_{n,h}} + \int_0^T\left(\|\mathbf{f}_n\|^2_{L^2(Q_n)^3} + \left\|\frac{\partial \mathbf{f}_n}{\partial t}\right\|^2_{L^2(Q_n)^3}\right)\mathrm{d}s\right). \tag{6.22}$$

Now, from (6.8), (6.2) and Cauchy Schwarz inequality, we get

$$\sigma^3(t_l)\|\eta^l_{n,h}\|^2_{L^2(Q_n)} \leq \sigma^3(t_l)\|d_t\mathbf{e}^l_{n,h}\|^2_{L^2(Q_n)^3} + \sigma^2(t_l)\|\mathbf{e}^l_{n,h}\|^2_1 + \sigma^3(t_l)\|E^l_{n,h}\|^2_{L^2(Q_n)^3}.$$

From (6.21), (6.18), (6.13), and (6.22) we get the thesis. □

From (6.19), (6.20), and (6.3), the following optimal a priori error estimate follows.

**Theorem 6.9** *Let $n$ be fixed. Let $(\mathbf{u}_n(t), p_n(t))$ be the solution of problem $(P_n)$ and its associated pressure, respectively, and let $(\mathbf{u}^l_{n,h}, p^l_{n,h})$ be the fully discretized solution of $(P^l_{n,h})$ and its associated pressure, respectively. Then for every $l = 0, 1, \ldots, N$ we have*

$$\sigma^2(t_l)\|\mathbf{u}_n(t_l) - \mathbf{u}^l_{n,h}\|^2_{X_{n,h}} + \sigma^3(t_l)\|p_n(t_l) - p^l_{n,h}\|^2_{L^2(Q_n)} \leq \|\mathbf{u}^0_n - \mathbf{u}^0_{n,h}\|^2_{X_{n,h}}$$

$$+ C_1 h^2\|\mathbf{f}_n\|^2_{L^2(Q_n)^3} + C_2\Delta t^2\int_0^T\left(\|\mathbf{f}_n\|^2_{L^2(Q_n)^3} + \left\|\frac{\partial \mathbf{f}_n}{\partial t}\right\|^2_{L^2(Q_n)^3} + \left\|\frac{\partial^2 \mathbf{f}_n}{\partial t^2}\right\|^2_{L^2(Q_n)^3}\right)\mathrm{d}s.$$

## 7 Numerical Simulations

In this section we present two numerical experiments on Stokes flows. We first consider a time-dependent problem, where a quiescent viscous fluid enclosed in a Koch-type pre-fractal pipe is accelerated by an external field. Secondly, we study stationary Stokes equations, and we aim to numerically validate the theoretical results on the mean shear stress along a fractal boundary, according to Sect. 5. The simulations have been performed on Comsol V.3.5, on a notebook computer with an Intel quad-Core i5-8250U processor running at 1.60 GHz and equipped with 16 GB RAM.

As to the time-dependent problem, we fix $n = 3$ and we consider problem $(P_n)$ as follows:

$$(P_n) \begin{cases} \rho \frac{\partial \mathbf{u}_n}{\partial t}(t,x) - \mu \Delta \mathbf{u}_n(t,x) + \nabla p_n(t,x) = \mathbf{f}_n(t,x) & \text{in } [0,T] \times Q_n, \\ \operatorname{div} \mathbf{u}_n(t,x) = 0 & \text{on } [0,T] \times Q_n, \\ \mathbf{u}_n(t,x) = 0 & \text{on } [0,T] \times \partial Q_n, \\ \mathbf{u}_n(0,x) = \mathbf{u}_n^0(x) & \text{in } Q_n, \end{cases}$$

where in this section $Q_n = \Omega_n \times [0,2]$, $\rho = 1$, $\mu = 0.1$, $\mathbf{f}_n = \left(0, 0, \sin(\pi x_1 \sqrt{3})\right)$, $T = 1$, $\Delta t = 0.01$ and $\mathbf{u}_n^0(x) = 0$.

The behavior of the third component of the velocity field for different time steps in the transient dynamic is described in Fig. 3. Since the external force is directed as the $x_3$ axis, far from the planes $x_3 = 0$ and $x_3 = 2$ (representing the inlet and the outlet of the fractal pipe) the relevant component of the velocity field is the one aligned in the same direction. On the other hand, in proximity of the bases $(\Omega_n \times \{0\}) \cup (\Omega_n \times \{2\})$, the velocity component $u_{1,n}$ starts increasing due to the boundary condition $\mathbf{u}_n(t,x) = 0$ on $\partial Q_n$. Thus the resulting motion is represented by an evolving vortex having as axis $x_2$.

Due to the dissipative nature of the Stokes operator, the velocity field quickly converges towards a stationary solution, as shown in the last panel of Fig. 3. In this case the fluid is quite "trapped" in the green areas of the domain as in Fig. 3.

As shown in Sect. 5, fractal-type boundaries can be useful in Hagen–Poiseuille-like flows, because they enhance the reduction of the mean shear stress exerted by the fluid onto the wall.

More precisely, in Hagen–Poiseuille-like flows, the equations of motion are represented by the stationary Stokes equations, where a pressure difference $[p_n] := p_{\perp,n}|_{\{x_3=0\}} - p_{\perp,n}|_{\{x_3=2\}}$ is imposed on the bases $\Omega_n \times \{0\}$ and $\Omega_n \times \{2\}$. In the stationary case, the pressure drop is balanced by the viscous shear stress exerted on the boundary. Due to the constant cross section, and the pressure condition on the bases, the resulting motion is unidirectional (hence it can be visualized in a generic section of the fractal pipe), and it is mathematically described by a scalar Poisson problem for the Laplace operator with homogeneous Dirichlet boundary conditions for the velocity component $u_{3,n}$. The values of $u_{3,n}$ in a cross section of the pipe are reported in Fig. 4. The different snapshots refer to different values of the parameter $n$.

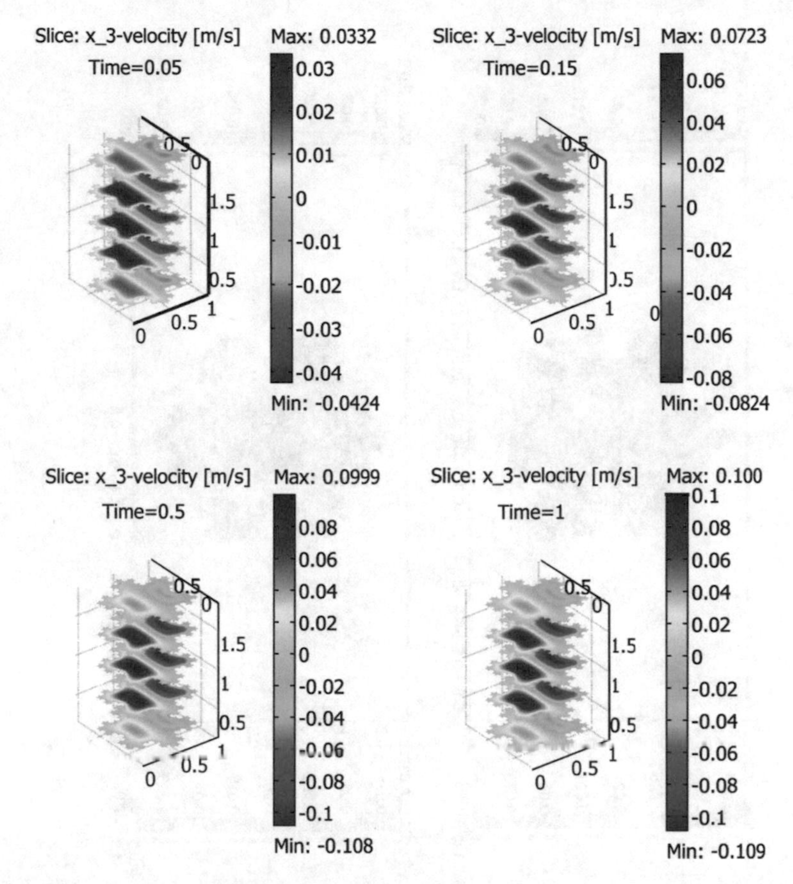


**Fig. 3** Time evolution of the velocity component $u_{3,n}$. The time sequence from the top (left panel) is referring to the time steps $t = 0.05,\ t = 0.15,\ t = 0.50$, and $t = 1.00$

It turns out that the maximum value of the velocity field is almost constant at each iteration. The specific form of the velocity profile is slightly influenced by the parameter $n$ and, as in the previous numerical example, a "quiescence zone" is detected towards the "trapping blue areas."

Looking at the mean value of the viscous shear stress (Table 1), we notice that $\langle \tau_{w,n} \rangle$ decreases as $n$ increases, according to Theorem 5.1.

Figure 5 shows the average wall shear stress data collected in Table 1. One may observe the exponential decay given by Theorem 5.1 (more precisely, in accordance with the curve $y = 0.22 \left(\frac{3}{4}\right)^n$). The reader may notice that the first two numerical values of Table 1 does not perfectly match with the exponential curve; this is due to numerical integration errors which are larger for coarse meshes, i.e. $n = 1, 2$.

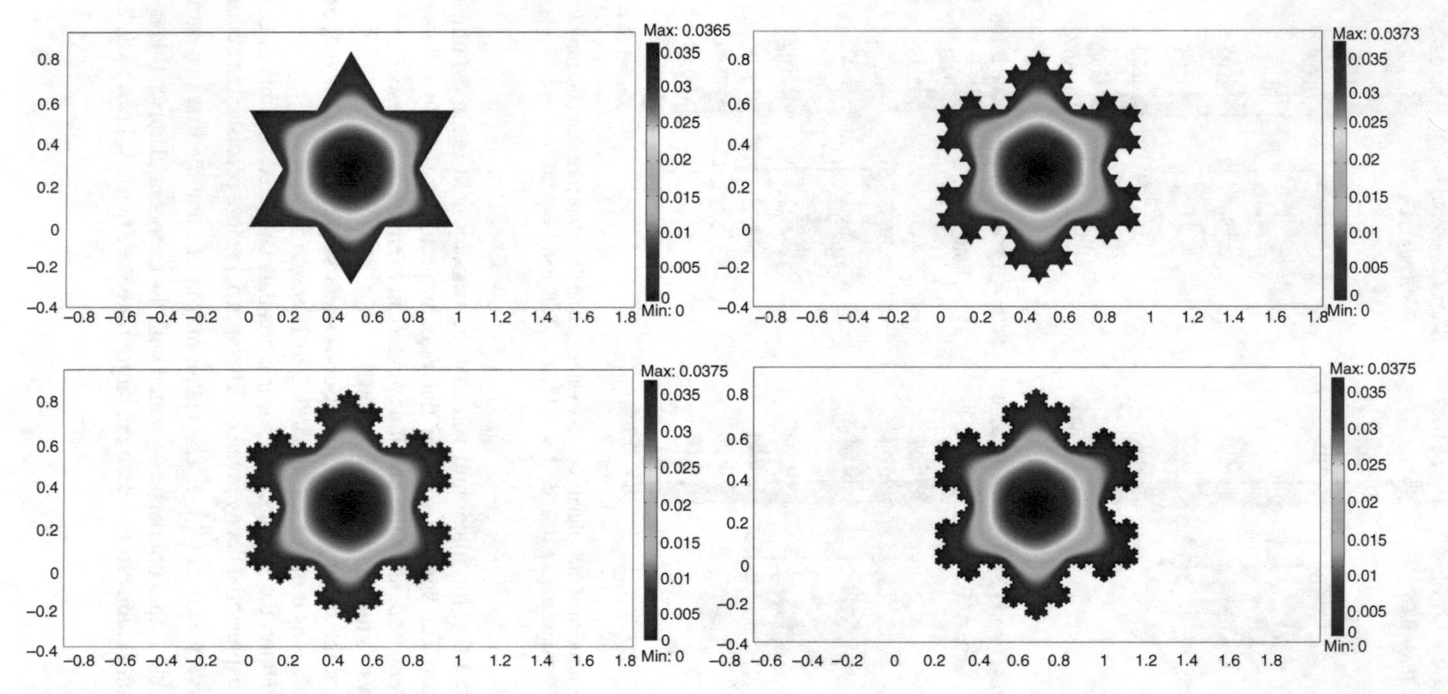


**Fig. 4** Snapshots of the velocity component $u_{3,n}$ for $n = 1, 3, 5, 7$

**Table 1** The values of the mean wall shear stress obtained in our simulations

| $n$ | $\langle\tau_{w,n}\rangle$ |
|---|---|
| 1 | 0.143352 |
| 2 | 0.118425 |
| 3 | 0.09157 |
| 4 | 0.069939 |
| 5 | 0.05179 |
| 6 | 0.038986 |
| 7 | 0.028286 |

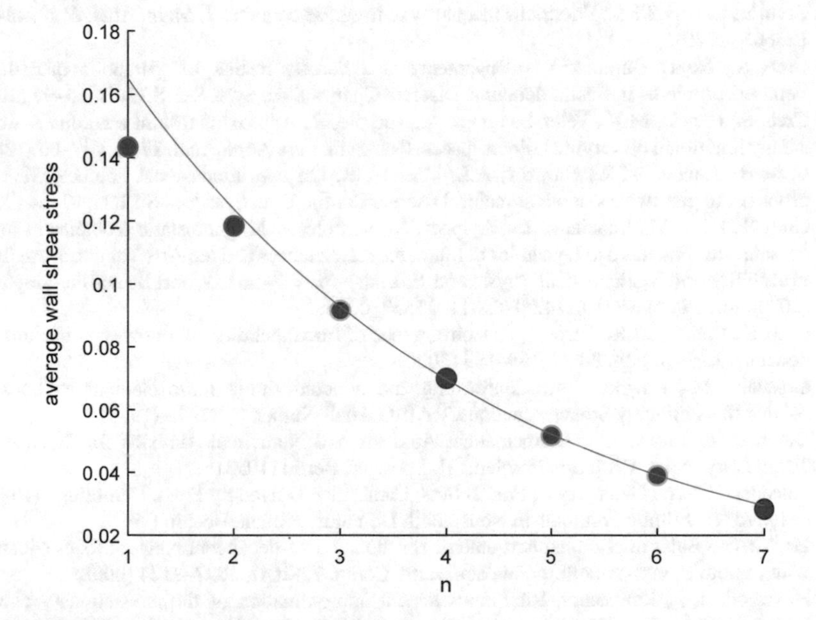


**Fig. 5** The exponential decay of $\langle\tau_{w,n}\rangle$

**Acknowledgments** S. C., M. R. L. and P. V. have been supported by the Gruppo Nazionale per l'Analisi Matematica, la Probabilità e le loro Applicazioni (GNAMPA) of the Istituto Nazionale di Alta Matematica (INdAM).

## References

1. Akkermans, E.: Statistical mechanics and quantum fields on fractals. In: Fractal Geometry and Dynamical Systems in Pure and Applied Mathematics. II, Fractals in Applied Mathematics. Contemp. Math., vol. 601, pp. 1–21. Amer. Math. Soc., Providence (2013)
2. Apel, T., Nicaise, S., Schöberl, J.: Crouzeix-Raviart type finite elements on anisotropic meshes. Numer. Math. **89**(2), 193–223 (2001)
3. Apel, T., Nicaise, S., Schöberl, J.: A non-conforming finite element method with anisotropic mesh grading for the Stokes problem in domains with edges. IMA J. Numer. Anal. **21**(4), 843–856 (2001)

4. Cefalo, M., Lancia, M.R.: An optimal mesh generation algorithm for domains with Koch type boundaries. Math. Comput. Simul. **106**, 133–162 (2014)
5. Cefalo, M., Dell'Acqua, G., Lancia, M.R.: Numerical approximation of transmission problems across Koch-type highly conductive layers. Appl. Math. Comput. **218**(9), 5453–5473 (2012)
6. Cefalo, M., Lancia, M.R., Liang, H.: Heat-flow problems across fractals mixtures: regularity results of the solutions and numerical approximation. Differ. Integr. Equ. **26**(9–10), 1027–1054 (2013)
7. Cefalo, M., Creo, S., Lancia, M.R., Vernole, P.: Nonlocal Venttsel' diffusion in fractal-type domains: regularity results and numerical approximation. Math. Methods Appl. Sci. **42**(14), 4712–4733 (2019)
8. Creo, S., Lancia, M.R.: Friedrichs inequality in irregular domains. J. Math. Anal. Appl. **484**(1), 123665 (2020)
9. Creo, S., Regis Durante, V.: Convergence and density results for parabolic quasi-linear Venttsel' problems in fractal domains. Discrete Contin. Dyn. Syst. Ser. S **12**(1), 65–90 (2019)
10. Creo, S., Lancia, M.R., Vélez-Santiago, A., Vernole, P.: Approximation of a nonlinear fractal energy functional on varying Hilbert spaces. Commun. Pure Appl. Anal. **17**(2), 647–669 (2018)
11. Creo, S., Lancia, M.R., Nazarov, A.I., Vernole, P.: On two-dimensional nonlocal Venttsel' problems in piecewise smooth domains. Discrete Contin. Dyn. Syst. Ser. S **12**(1), 57–64 (2019)
12. Creo, S., Hinz, M., Lancia, M.R., Teplyaev, A., Vernole, P.: Magnetostatic problems in fractal domains. In: Fractals and Dynamics in Mathematics, Sciences and the Arts Volume 5: Analysis, Probability and Mathematical Physics on Fractals, pp. 477–502. World Scientific, Singapore (2020). https://doi.org/10.1142/9789811215537_0015.
13. Creo, S., Lancia, M.R., Vernole, P.: Convergence of fractional diffusion processes in extension domains. J. Evol. Equ. **20**(1), 109–139 (2020)
14. Crouzeix, M., Raviart, P.A.: Conforming and nonconforming finite element methods for solving the stationary Stokes equations. RAIRO Anal. Numér. **7**, 33–75 (1973)
15. Dautray, R., Lions, J.L.: Mathematical Analysis and Numerical Methods for Science and Technology, vol 6. Evolution Problems II. Springer, Berlin (1993)
16. Falconer, K.: The Geometry of Fractal Sets. Cambridge University Press, Cambridge (1986)
17. Grisvard, P.: Elliptic Problems in Nonsmooth Domains. Pitman, Boston (1985)
18. He, Y.: The Euler implicit/explicit scheme for the 2D time-dependent Navier-Stokes equations with smooth or non-smooth initial data. Math. Comp. **77**(264), 2097–2124 (2008)
19. Heywood, J.G., Rannacher, R.: Finite-element approximation of the nonstationary Navier-Stokes problem, Part IV: error analysis for second-order time discretization. SIAM J. Numer. Anal. **27**(2), 353–384 (1990)
20. Hinz, M., Rogers, L.: Magnetic fields on resistance spaces. J. Fractal Geom. **3**(1), 75–93 (2016)
21. Hinz, M., Teplyaev, A.: Dirac and magnetic Schrödinger operators on fractals. J. Funct. Anal. **265**(11), 2830–2854 (2013)
22. Jerison, D., Kenig, C.E.: The inhomogeneous Dirichlet problem in Lipschitz domains. J. Funct. Anal. **130**(1), 161–219 (1995)
23. Jonsson, A., Wallin, H.: Function Spaces on Subsets of $\mathbb{R}^n$, Part 1, Math. Reports, vol. 2. Harwood Acad. Publ., London (1984)
24. Kato, T.: Perturbation Theory for Linear Operators, II Edit. Springer, Berlin (1976)
25. Lancia, M.R., Vernole, P.: Venttsel' problems in fractal domains. J. Evol. Equ. **14**(3), 681–712 (2014)
26. Lancia, M.R., Vernole, P.: The Stokes problems in fractal domains: asymptotic behaviour of the solutions. Discrete Contin. Dyn. Syst. Ser. S **13**(5), 1553–1565 (2020)
27. Lancia, M.R., Vélez-Santiago, A., Vernole, P.: Quasi-linear Venttsel' problems with nonlocal boundary conditions on fractal domains. Nonlinear Anal. Real World Appl. **35**, 265–291 (2017)
28. Monniaux, S.: Navier-Stokes equations in arbitrary domains: the Fujita-Kato scheme. Math. Res. Lett. **13**, 455–461 (2006)
29. Nicaise, S.: Regularity of the solutions of elliptic systems in polyhedral domains. Bull. Belg. Math. Soc. Simon Stevin **4**(3), 411–429 (1997)

30. Quarteroni, A., Valli, A.: Numerical Approximation of Partial Differential Equations. Springer, Berlin (1994)
31. Shen, S., Xu, J., Zhou, J., Chen, Y.: Flow and heat transfer in microchannels with rough wall surface. Energy Convers. Manage. **47**, 1311–1325 (2006)
32. Sohr, H.: The Navier-Stokes Equations. An Elementary Functional Analytic Approach. Birkhäuser Verlag, Basel (2001)
33. Stokes, G.G.: On the theories of the internal friction of fluids in motion, and of the equilibrium and motion of elastic solids. Trans. Camb. Philos. Soc. **8**, 287–341 (1845)
34. Sutera, S.P., Skalak, R.: The history of Poiseuille's law. Ann. Rev. Fluid Mech. **25**, 1–19 (1993). Annual Reviews, Palo Alto, CA
35. Taylor, B., Carrano, A.L., Kandlikar, S.G.: Characterization of the effect of surface roughness and texture on fluid flow past, present, and future. Int. J. Thermal Sci. **45**, 962–968 (2006)
36. Temam, R.: Roger Navier-Stokes Equations. Theory and Numerical Analysis. Studies in Mathematics and Its Applications, vol. 2. North-Holland, Amsterdam (1979)
37. Yang, S.S., Yu, B., Zou, M., Liang, M.: A fractal analysis of laminar flow resistance in roughened microchannels. Int. J. Heat Mass Transf. **77**, 208–217 (2014)

# $\infty$-Laplacian Obstacle Problems in Fractal Domains

**Salvatore Fragapane**

**Abstract** In this paper we deal with $p$-Laplacian and $\infty$-Laplacian obstacle problems in fractal and pre-fractal domains, analyzing both the asymptotic behavior and the issue of the uniqueness of the solutions.

Moreover, we consider numerical approximations, we state uniform estimates for FEM-approximate solutions and we discuss about the rate of vanishing of the approximation error.

**Keywords** Fractals · Degenerate elliptic equations · Obstacle problems · Asymptotic behavior · FEM

## 1 Introduction

In this paper we will study obstacle problems involving both $p$-Laplacian and the limit operator $\infty$-Laplacian in fractal and pre-fractal boundary domains.

The study of fractal and pre-fractal boundary domains, and more in general of fractal and pre-fractal sets, finds its motivations in the possibility of these objects to describe Nature in a new and more appropriate way opposite to the classical shapes of Geometry (the pulmonary alveolus is an example). Fractals are powerful tools in the construction of models for various phenomena in many fields: Biology, Medicine, Engineering, etc. (see, for instance, [14] and [15]). For example, they are an excellent resource for the description of all phenomena in which surface effects are prevalent on volume ones.

---

The author is member of GNAMPA (INdAM).

S. Fragapane (✉)
Dipartimento di Scienze di Base e Applicate per l'Ingegneria, "Sapienza" Università di Roma, Roma, Italy
e-mail: salvatore.fragapane@uniroma1.it; salvatore.fragapane@sbai.uniroma1.it



M. R. Lancia, A. Rozanova-Pierrat (eds.), *Fractals in Engineering: Theoretical Aspects and Numerical Approximations*, SEMA SIMAI Springer Series 8,
https://doi.org/10.1007/978-3-030-61803-2_3

The choice of $p$-Laplace type operators is due to their massive use in the study of many physical processes: quasi-Newtonian flow, power law material, non-linear filtration and diffusion (see [6] and the references therein). Moreover, the solutions of problems involving $\infty$-Laplacian provide an answer for the mass transport problem (see, f.i., [8] and [16]) and even for the problem of torsion creep (see, for instance, [12] and [1] and the references therein).

In the present paper, for $p \in (2, \infty)$, we will consider the following obstacle problems:

$$\text{find } u_p \in \mathcal{K}_p \ : \ J_p(u_p) = \min_{v \in \mathcal{K}_p} J_p(v), \ \text{with } J_p(v) = \frac{1}{p} \int_{\Omega_\alpha} |\nabla v|^p \mathrm{d}x - \int_{\Omega_\alpha} f v \mathrm{d}x, \tag{1.1}$$

and

$$\text{find } u_\infty \in \mathcal{K}^\infty \ : \ J(u_\infty) = \max_{w \in \mathcal{K}^\infty} J(w), \text{ with } J(w) = \int_{\Omega_\alpha} w(x) f(x)\, \mathrm{d}x, \tag{1.2}$$

where $\mathcal{K}_p$ and $\mathcal{K}^\infty$ are convex subsets of $W^{1,p}(\Omega_\alpha)$ and $W^{1,\infty}(\Omega_\alpha)$, respectively. The datum $f \in L^1(\Omega_\alpha)$ and the obstacle $\varphi \in C(\overline{\Omega}_\alpha)$ are given.

As in [2], for the case of two obstacle, we will show that the sequence of solutions of Problems (1.1), as $p \to \infty$, converges to a solution of Problem (1.2). Moreover, we will consider the analogous problems in the pre-fractal approximating domains $\Omega_\alpha^n$, providing an analogous result to the one given in the fractal case.

Along with these first convergence results, we will show that the sequence of solutions of pre-fractal problems (i.e. the problems on the approximating domains $\Omega_\alpha^n$) converges, as $n \to \infty$, to a solution of the problem on the corresponding fractal boundary domain for $p = \infty$.

Furthermore, we will prove that it is sufficient to require that the final convex $\mathcal{K}^\infty$ is non-empty to ensure that both the pre-fractal problems and the fractal ones admit solutions. More precisely, to assume $\mathcal{K}^\infty \neq \emptyset$ will allow us to construct a sequence of continuous obstacle $\varphi_n$ converging to $\varphi$ in $C(\overline{\Omega}_\alpha)$ such that the corresponding convex sets $\mathcal{K}_n^\infty$ will be non-empty (and, a fortiori, we have $\mathcal{K}_p \neq \emptyset$ and $\mathcal{K}_{p,n} \neq \emptyset$).

In addition to the asymptotic behavior of the solutions, we will present uniqueness results, which are an improvement of Theorems 4.3 and 4.5 in [4]. In fact, here we obtain the same conclusion requiring weaker assumptions; moreover, we obtain the explicit form of the solution.

Finally, we will state the existence of the solutions of the approximate problems with the Galerkin method, that is the solutions of FEM-problems. Moreover, we will prove, for the FEM-solutions, uniform estimates with respect to $p$ and $h$ (the size of the triangulation). We will also discuss about the error estimates (we refer to [3] and [5] for the case of double obstacle). As intermediate results, we will show that the non-emptiness of $\mathcal{K}^\infty$ guarantees the non-emptiness of the convex $\mathcal{K}_{n,h}$ in which we search the FEM-solution. In the framework of numerical analysis, we want to stress the fact that in the case of uniqueness of the solution the approximate solutions $u^h_{p,n}$ give an approximation of $u_p$, $u_{\infty,n}$ and $u_\infty$.

All these results are resumed in the scheme in Fig. 6, which provide an improvement with respect to the results summarized in the scheme in Fig. 4 and a completion of the ones represented in Fig. 5.

The organization of the paper is the following. In Sect. 2 definitions and properties regarding the sets $\Omega^n_\alpha$ and $\Omega_\alpha$ and suitable fibers $\Sigma^n$ are recalled. In Sect. 3 the problems are introduced and results about asymptotic behavior of solutions both as $p \to \infty$ and $n \to \infty$ are given. In Sect. 4 uniqueness results are provided. Finally, in Sect. 5 uniform estimates for the solutions of the corresponding FEM-problems are proven and the issue of the rate of vanishing of the approximation error is discussed.

## 2 Fractal Domains, Approximating Domains and Fibers

In the following sections we will consider both Koch Island, i.e. domains $\Omega_\alpha$ having Koch curve as boundary, and polygonal domains $\Omega^n_\alpha$ which approximate them (see Fig. 1).

To construct pre-fractal approximating domains $\Omega^n_\alpha$ we start from any regular polygon $\Omega^0$ (triangle, square, pentagon, etc.) and we replace each side with the n-th pre-fractal Koch curve $K^n_\alpha$ (see Fig. 2).

To obtain $K^n_\alpha$, we proceed as follow (see [10] for details and proofs).

Let us consider the line segment $K^0$ with endpoints $P(0, 0)$ and $Q(1, 0)$, for instance, and let us introduce a family $\Psi_\alpha = \{\psi_{1,\alpha}, \dots, \psi_{4,\alpha}\}$ of four contractive similarities having as contraction factor $\alpha^{-1}$, with $2 < \alpha < 4$, defined as follow:

$$\psi_{1,\alpha}(z) = \frac{z}{\alpha}, \qquad \psi_{2,\alpha}(z) = \frac{z}{\alpha}e^{i\theta(\alpha)} + \frac{1}{\alpha},$$

$$\psi_{3,\alpha}(z) = \frac{z}{\alpha}e^{-i\theta(\alpha)} + \frac{1}{2} + i\sqrt{\frac{1}{\alpha} - \frac{1}{4}}, \qquad \psi_{4,\alpha}(z) = \frac{z-1}{\alpha} + 1,$$

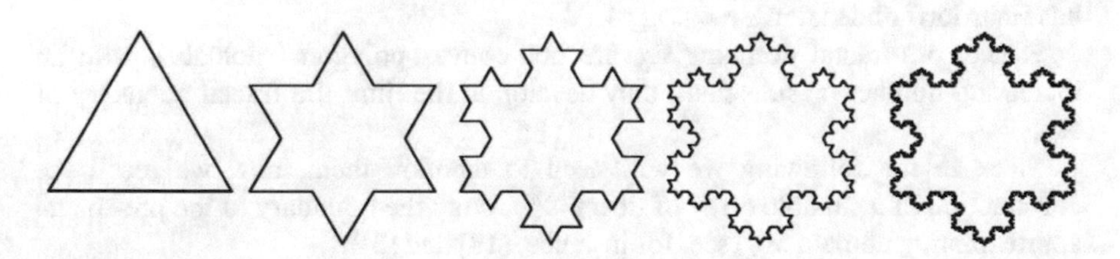

**Fig. 1** $\Omega^0$, $\Omega^n_3$, for $n = 1, 2, 3$, and $\Omega_\alpha$

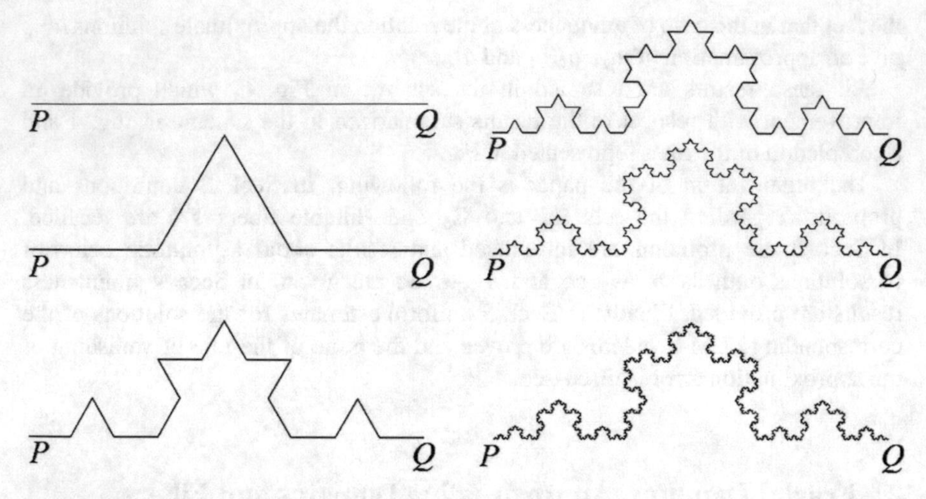


**Fig. 2** $K^0$, $K_3^n$ for $n = 1, 2, 3, 4$, and $K_3$

where

$$\theta(\alpha) = \arcsin\left(\frac{\sqrt{\alpha(4-\alpha)}}{2}\right). \tag{2.1}$$

Thus, for each $n \in \mathbb{N}$, we set

$$K_\alpha^n = \bigcup_{i=1}^{4} \psi_{i,\alpha}(K_\alpha^{n-1}) = \bigcup_{i|n} K_\alpha^{i|n}, \text{ with } K_\alpha^{i|n} = \psi_{i|n,\alpha}(K^0),$$

where $\psi_{i|n,\alpha} = \psi_{i_1,\alpha} \circ \psi_{i_2,\alpha} \circ \cdots \circ \psi_{i_n,\alpha}$, is the identity map in $\mathbb{R}^2$ for $n = 0$ and it is the map associated with an arbitrary $n$-tuple of indices $i|n = (i_1, i_2, \ldots, i_n) \in \{1, \ldots, 4\}^n$ for each integer $n > 0$.

For each $n \in \mathbb{N}$, the set $K_\alpha^n$ is the so-called $n$-th pre-fractal Koch curve.

For $n \to \infty$ the curves $K_\alpha^n$ converge to the fractal curve $K_\alpha$ in the Hausdorff metric. Moreover, $K_\alpha$ it is the unique compact set which is invariant on $\Psi_\alpha$ and it has Hausdorff dimension $d_f = \log_\alpha 4$.

Hence, pre-fractal domains $\Omega_\alpha^n$ are non-convex polygonal domains, with an increasing number of sides, and they develop at the limit the fractal geometry of $\Omega_\alpha$.

Since in the following we will need to mention them, now we recall the construction of a suitable array of fibers $\Sigma^n$ around the boundary of the pre-fractal approximating domain $\Omega_\alpha^n$ (see, for instance, [18] and [3]).

In particular, we will use the internal fibers; nevertheless, for sake of completeness, we will show the whole construction.

In order to understand how construction works, let us start from the open triangle of vertices $P(0,0)$, $Q(1,0)$ and $R(1/2, -\sqrt{3}/2)$. By $T_0^+$ we denote the open triangle of vertices $P = (0,0)$, $Q(1,0)$ and $S^+(1/2, \delta_+/2)$, where $\delta_+ = \tan(\frac{\vartheta}{2})$ and $\vartheta$ is the rotation angle (2.1). The triangle $T_0^+$ satisfies the open set condition with respect to the maps $\Psi_\alpha$, this means that $\psi_{i|n,\alpha}(T_0^+) \subset T_0^+$ for every $i|n$ and $\psi_{i|n,\alpha}(T_0^+) \cap \psi_{j|n,\alpha}(T_0^+) = \emptyset$ for every $i|n \neq j|n$. Moreover, by $T_0^-$ we denote the open triangle of vertices $P(0,0)$, $Q(1,0)$ and $S^-(1/2, -\delta_-/2)$, where $\delta_- = \tan(\vartheta^-)$, with $0 < \vartheta^- \leq \min\{\pi/2 - \vartheta, \vartheta/2\}$. So, we obtain the fiber $\Sigma_1^0$ corresponding to the side $PQ$ setting

$$\Sigma_1^0 = T_0^+ \bigcup T_0^- \bigcup K^0.$$

At this point, we iteratively transform $\Sigma_1^0$ into increasingly fine arrays, applying the maps $\psi_{i|n} = \psi_{i_1} \circ \psi_{i_2} \circ \cdots \circ \psi_{i_n}$, for any integer $n > 0$. In particular, for every $n \geq 1$, we define

$$\Sigma_1^n = \Sigma_{1,+}^n \bigcup \Sigma_{1,-}^n \bigcup K^n,$$

where

$$\Sigma_{1,+}^n = \bigcup_{i|n} \Sigma_{1,+}^{i|n}, \quad \Sigma_{1,+}^{i|n} = \psi_{i|n}(T_0^+), \tag{2.2}$$

$$\Sigma_{1,-}^n = \bigcup_{i|n} \Sigma_{1,-}^{i|n}, \quad \Sigma_{1,-}^{i|n} = \psi_{i|n}(T_0^-). \tag{2.3}$$

Repeating the same steps procedure for the others sides $PR$ and $QR$ of the initial domain and denoting by $\Sigma_{2,+}^n$, $\Sigma_{3,+}^n$, $\Sigma_{2,-}^n$ and $\Sigma_{3,-}^n$ the corresponding arrays of fiber, we get

$$\Sigma^n = \bigcup_{j=1,2,3} \Sigma_j^n, \qquad \Sigma_+^n = \bigcup_{j=1,2,3} \Sigma_{j,+}^n, \qquad \Sigma_-^n = \bigcup_{j=1,2,3} \Sigma_{j,-}^n. \tag{2.4}$$

Hence, we set

$$\hat{\Omega}_\alpha^n = int(\bar{\Omega}_\alpha^n \bigcup \Sigma_+^n) \quad \text{and} \quad \breve{\Omega}_\alpha^n = \Omega_\alpha^n \setminus \overline{\Sigma}_-^n. \tag{2.5}$$

The following inclusions hold

$$\breve{\Omega}_\alpha^n \subset \Omega_\alpha^n \subset \hat{\Omega}_\alpha^n \tag{2.6}$$

and

$$\hat{\Omega}_\alpha^{n+1} \subset \hat{\Omega}_\alpha^n \qquad \text{and} \qquad \breve{\Omega}_\alpha^n \subset \breve{\Omega}_\alpha^{n+1}. \tag{2.7}$$

**Fig. 3** $\overline{\Omega}_3^n$ as union of $\breve{\Omega}_3^n$ (in green) and $\overline{\Sigma}_-^n$ (in yellow), for $n = 0$ and $n = 1$

In Fig. 3 we see some examples of the domains $\Omega_\alpha^n$ join to the internal part of the array of fibers.

## 3 Setting and Asymptotic Behavior

Let us consider $p \in (2, \infty)$. Given $f \in L^1(\Omega_\alpha)$ and a function $\varphi \in C(\overline{\Omega}_\alpha)$, let us consider the following obstacle problems:

$$\text{find } u_p \in \mathcal{K}_p \ : \ \int_{\Omega_\alpha} |\nabla u_p|^{p-2}\nabla u_p \nabla(v - u_p)\mathrm{d}x - \int_{\Omega_\alpha} f(v - u_p)\mathrm{d}x \geqslant 0,\ \forall v \in \mathcal{K}_p, \tag{$\mathcal{P}_p$}$$

with

$$\mathcal{K}_p = \{v \in W_0^{1,p}(\Omega_\alpha) : v \geqslant \varphi \text{ in } \Omega_\alpha\} \neq \emptyset. \tag{3.1}$$

*Remark 3.1* Problem $(\mathcal{P}_p)$ is equivalent to the following minimum problem

$$\min_{v \in \mathcal{K}_p} J_p(v), \text{ with } J_p(v) = \frac{1}{p}\int_{\Omega_\alpha} |\nabla v|^p \,\mathrm{d}x - \int_{\Omega_\alpha} f v \,\mathrm{d}x. \tag{3.2}$$

Furthermore (see, for instance, [20]), since $\mathcal{K}_p$ is a non-empty, closed and convex subset of $W^{1,p}(\Omega)$ and the functional $J_p(v)$ is convex, weakly lower semi-continuous and coercive, then Problem (3.2) has a minimizer $u_p$ in $\mathcal{K}_p$.

From now on we will use the following notation:

$$||u||_{L^p(\Omega_\alpha)} = ||u||_p \text{ and } ||u||_{L^\infty(\Omega_\alpha)} = ||u||_\infty,$$

$$||u||_{L^p(\Omega_\alpha^n)} = ||u||_{p,n} \text{ and } ||u||_{L^\infty(\Omega_\alpha^n)} = ||u||_{\infty,n},$$

$$||u||_{W^{1,p}(\Omega_\alpha)} = ||u||_{1,p} \text{ and } ||u||_{W^{s,r}(\Omega_\alpha^n)} = ||u||_{s,r,n}.$$

Since we will use it in the following, let us recall the following lemma (see, for instance, the proof of Theorem 4 in Section 5.6.2 in [7]).

**Lemma 3.1** *Let $\Omega$ be an open subset of $\mathbb{R}^2$, $p > 2$ and $u \in W_0^{1,p}(\Omega)$. Then*

$$\sup_{x\in\Omega} |u| \leq \pi^{-\frac{1}{p}}||u||_p + (2\pi)^{-\frac{1}{p}}\left(\frac{p-1}{p-2}\right)^{1-\frac{1}{p}}||\nabla u||_p.$$

Moreover, see Theorem 12.17 in [13], we recall the

**Proposition 3.1 (Poincaré's Inequality)** *Let $\Omega$ be a bounded open subset of $\mathbb{R}^2$, $p \geq 1$ and $u \in W_0^{1,p}(\Omega)$. Then*

$$||u||_p \leq \frac{d}{p^{\frac{1}{p}}}||\nabla u||_p \leq d||\nabla u||_p, \tag{3.3}$$

*where $d$ is the diameter of $\Omega$.*

*Remark 3.2* Since we have (homogeneous) Dirichlet condition, Problem $(\mathcal{P}_p)$ admits at most a solution $u_p$.

Let us consider the following problem:

$$\text{find } u_\infty \in \mathcal{K}^\infty \ : \int_{\Omega_\alpha} u_\infty(x) f(x)\,\mathrm{d}x = \max\left\{\int_{\Omega_\alpha} w(x) f(x)\,\mathrm{d}x \ : \ w \in \mathcal{K}^\infty\right\}, \tag{$\mathcal{P}$}$$

where

$$\mathcal{K}^\infty = \{v \in W_0^{1,\infty}(\Omega_\alpha) \ : \ v \geqslant \varphi \text{ in } \Omega_\alpha,\ ||\nabla v||_\infty \leq 1\}, \tag{3.4}$$

with $W_0^{1,\infty}(\Omega_\alpha) := \{w \in W^{1,\infty}(\Omega_\alpha) \ : \ w = 0 \text{ on } \partial\Omega_\alpha\}$.

The following result states the asymptotic behavior of solution $u_p$ for $p \to \infty$ (see also [1, 16] and [2]).

We underline that in the following we will use $C$ to indicate possibly different constants.

**Theorem 3.1** *Let us assume $f \in L^1(\Omega_\alpha)$, $\mathcal{K}^\infty \neq \emptyset$ and, for any $p$, let $u_p$ be the minimizer of Problem $(\mathcal{P}_p)$. Then the family of minimizer $\{u_p\}_{p>m}$ is pre-compact in $C(\overline{\Omega}_\alpha)$. In particular, for any subsequence $u_{p_k}$ there exists a subsubsequence still denoted with $u_{p_k}$ such that, as $k \longrightarrow \infty$, $u_{p_k} \longrightarrow u_\infty$ weakly in $W^{1,r}(\Omega_\alpha)$, being $u_\infty$ a maximizer of Problem $(\mathcal{P})$.*

***Proof*** For any function $v \in \mathcal{K}^\infty$, by Remark 3.1, we have

$$\frac{1}{p}||\nabla u_p||_p^p = \frac{1}{p}\int_{\Omega_\alpha} |\nabla u_p|^p \, dx \le \frac{1}{p}\int_{\Omega_\alpha} |\nabla v|^p \, dx - \int_{\Omega_\alpha} f v \, dx + \int_{\Omega_\alpha} f u_p \, dx.$$

Thanks to Hölder's inequality and Lemma 3.1, we obtain

$$\frac{1}{p}||\nabla u_p||_p^p \le \frac{|\Omega_\alpha|}{p} + ||f||_1||v||_\infty + ||f||_1\Big[(\pi)^{-\frac{1}{p}}||u||_p + (2\pi)^{-\frac{1}{p}}\Big(\frac{p-1}{p-2}\Big)^{1-\frac{1}{p}}||\nabla u||_p\Big].$$

Now, applying relation (3.3) and Young's inequality, with conjugate exponents $p$ and $p'$, we get

$$\frac{1}{p}||\nabla u_p||_p^p \le \frac{|\Omega_\alpha|}{p} + ||f||_1||v||_\infty + \frac{\varepsilon^{-\frac{1}{p-1}}||f||_1^{p'}(\pi)^{-\frac{1}{p-1}}d^{p'}}{p' p^{\frac{p'}{p}}} + \frac{\varepsilon||\nabla u_p||_p^p}{p} + \frac{\varepsilon^{-\frac{1}{p-1}}||f||_1^{p'}(2\pi)^{-\frac{1}{p-1}}\left(\frac{p-1}{p-2}\right)}{p'} + \frac{\varepsilon||\nabla u_p||_p^p}{p}.$$

So, by choosing $\varepsilon = \frac{1}{4}$, we get

$$\frac{1}{p}||\nabla u_p||_p^p \le \frac{2|\Omega_\alpha|}{p} + 2||f||_1||v||_\infty + \frac{(p-1)2^{\frac{p+1}{p-1}}||f||_1^{\frac{p}{p-1}}\pi^{-\frac{1}{p-1}}d^{\frac{p}{p-1}}}{p^{p'}} + \frac{(p-1)2^{\frac{p}{p-1}}||f||_1^{\frac{p}{p-1}}(\pi)^{-\frac{1}{p-1}}\left(\frac{p-1}{p-2}\right)}{p}.$$

Then

$$||\nabla u_p||_p^p \le 2|\Omega_\alpha| + 2p||f||_1||v||_\infty + \frac{(p-1)2^{\frac{p+1}{p-1}}||f||_1^{\frac{p}{p-1}}\pi^{-\frac{1}{p-1}}d^{\frac{p}{p-1}}}{p^{\frac{1}{p-1}}} + (p-1)2^{\frac{p}{p-1}}||f||_1^{\frac{p}{p-1}}(\pi)^{-\frac{1}{p-1}}\Big(\frac{p-1}{p-2}\Big).$$

Now, considering $m > 2$ and $p > m$ and setting

$$C = \max\Big\{2|\Omega_\alpha|,\ 2||f||_1||v||_\infty,\ \sup_{p>m}\Big\{\frac{2^{\frac{p+1}{p-1}}||f||_1^{\frac{p}{p-1}}\pi^{-\frac{1}{p-1}}d^{\frac{p}{p-1}}}{p^{\frac{1}{p-1}}}\Big\},$$
$$\sup_{p>m}\Big\{2^{\frac{p}{p-1}}||f||_1^{\frac{p}{p-1}}\pi^{-\frac{1}{p-1}}\Big(\frac{p-1}{p-2}\Big)\Big\}\Big\},$$

we have

$$||\nabla u_p||_p \leq ((3p+1)C)^{\frac{1}{p}}. \tag{3.5}$$

Thus, thanks again to relation (3.3), we obtain that the family $u_p$ is bounded in $W^{1,p}(\Omega_\alpha)$.

So, we have

$$||\nabla u_p||_m^m \leq ||\nabla u_p||_p^m |\Omega_\alpha|^{1-\frac{m}{p}} \leq ((3p+1)C)^{\frac{m}{p}}|\Omega_\alpha|^{1-\frac{m}{p}}.$$

Thus

$$\limsup_{p\to\infty} ||\nabla u_p||_m \leq |\Omega_\alpha|^{\frac{1}{m}}. \tag{3.6}$$

Then, for (3.6), thanks to Morrey-Sobolev's embedding, we have

$$|u_p(x) - u_p(y)| \leq C(m)|x-y|^{1-\frac{2}{m}}. \tag{3.7}$$

By Ascoli–Arzelà compactness criterion, by using (3.7) we can extract a subsequence of the previous one, that we indicate again with $\{u_p\}$, such that

$$u_p \longrightarrow u_\infty \text{ uniformly in } \overline{\Omega}_\alpha,\ \text{ for } p \longrightarrow \infty. \tag{3.8}$$

Moreover, thanks to Lemma 3.1 and (3.6) we have, for all $m > 2$,

$$||u_p||_{1,m} \leq C$$

(with $C$ independent from $p$). Then, there exists a subsequence denoted by $u_{p_k}$, such that, for $k \longrightarrow \infty$,

$$u_{p_k} \longrightarrow u_m^* \text{ weakly in } W^{1,m}(\Omega_\alpha).$$

From (3.8), we deduce that $u_m^* = u_\infty$ and then the whole sequence $\{u_p\}$ in (3.8) converges in $W^{1,m}(\Omega_\alpha)$ to $u_\infty$.

Thus

$$||\nabla u_\infty||_m \leq \liminf_{p\to\infty} ||\nabla u_p||_m,$$

then, passing to the limit as $m \to \infty$ (see [19, Chapter 6]), by (3.6), we obtain

$$||\nabla u_\infty||_\infty = \lim_{m\to\infty} ||\nabla u_\infty||_m \leq \lim_{m\to\infty} \liminf_{p\to\infty} ||\nabla u_p||_m \leq 1,$$

so we obtain that $u_\infty \in \mathcal{K}^\infty$.

Finally, since for any function $w \in \mathcal{K}^\infty$, we have

$$-\int_{\Omega_\alpha} f u_p \,\mathrm{d}x \leq \frac{1}{p}\int_{\Omega_\alpha} |\nabla u|^p \,\mathrm{d}x - \int_{\Omega_\alpha} f u_p \,\mathrm{d}x \leq \frac{|\Omega_\alpha|}{p} - \int_{\Omega_\alpha} f w \,\mathrm{d}x,$$

then, by (3.6), passing to the limit as $p \to \infty$, we obtain

$$\lim_{p\to\infty}\left(-\int_{\Omega_\alpha} f u_p \,\mathrm{d}x\right) \leq \lim_{p\to\infty}\left(\frac{|\Omega_\alpha|}{p} - \int_{\Omega_\alpha} f w \,\mathrm{d}x\right),$$

so

$$-\int_{\Omega_\alpha} f u_\infty \,\mathrm{d}x \leq -\int_{\Omega_\alpha} f w \,\mathrm{d}x,$$

that is

$$\int_{\Omega_\alpha} f u_\infty \,\mathrm{d}x \geq \int_{\Omega_\alpha} f w \,\mathrm{d}x.$$

Thus, we get

$$\int_{\Omega_\alpha} u_\infty(x) f(x) \,\mathrm{d}x = \max\left\{\int_{\Omega_\alpha} w(x) f(x) \,\mathrm{d}x : w \in \mathcal{K}^\infty\right\},$$

which conclude the proof. □

Along with the problems in the fractal boundary domain $\Omega_\alpha$, we can also consider the analogous problems in the pre-fractal approximating domains $\Omega_\alpha^n$. This becomes necessary, for instance, in order to perform numerical analysis, as we will do next.

In particular, let $n$ be a fixed positive integer and $p \in (2, \infty)$. Given $f \in L^1(\Omega_\alpha)$ and a function $\varphi_n \in C(\overline{\Omega_\alpha^n})$, let us consider the following obstacle problems

$$\text{find } u_{p,n} \in \mathcal{K}_{p,n} : \int_{\Omega_\alpha^n} |\nabla u_{p,n}|^{p-2} \nabla u_{p,n} \nabla(v - u_{p,n}) \,\mathrm{d}x - \int_{\Omega_\alpha^n} f(v - u_{p,n}) \,\mathrm{d}x \geqslant 0,\ \forall v \in \mathcal{K}_{p,n}, \tag{$\mathcal{P}_{p,n}$}$$

with

$$\mathcal{K}_{p,n} = \{v \in W_0^{1,p}(\Omega_\alpha^n) : v \geqslant \varphi_n \text{ in } \Omega_\alpha^n\} \neq \emptyset. \tag{3.9}$$

Again thanks to Remarks 3.1 and 3.2, which still hold replacing $\Omega_\alpha$ with $\Omega_\alpha^n$, we have that Problems $(\mathcal{P}_{p,n})$ admit a unique solution for any $n \in \mathbb{N}$.

Analogously to before, we consider the problem

$$\int_{\Omega_\alpha^n} u_{\infty,n}(x) f(x)\, \mathrm{d}x = \max\left\{\int_{\Omega_\alpha^n} w(x) f(x)\, \mathrm{d}x \ : \ w \in \mathcal{K}_n^\infty\right\}, \tag{$\mathcal{P}_n$}$$

where

$$\mathcal{K}_n^\infty = \{v \in W_0^{1,\infty}(\Omega_\alpha^n) \ : \ v \geqslant \varphi_n \text{ in } \Omega_\alpha^n, \ ||\nabla v||_{\infty,n} \leq 1\}, \tag{3.10}$$

with $W_0^{1,\infty}(\Omega_\alpha^n) := \{w \in W^{1,\infty}(\Omega_\alpha^n) \ : \ w = 0 \text{ on } \partial\Omega_\alpha^n\}$.

In an analogous way, we can prove that the following result for the asymptotic behavior of solution of Problems $(\mathcal{P}_{p,n})$ holds.

**Theorem 3.2** *Let us assume $f \in L^1(\Omega_\alpha)$, $\mathcal{K}_n^\infty \neq \emptyset$ and, for any $p$, let $u_{p,n}$ be the minimizer of Problem $(\mathcal{P}_{p,n})$. Then the family of minimizer $\{u_{p,n}\}_{p>m}$ is pre-compact in $C(\overline{\Omega}_\alpha^n)$. In particular, for any subsequence $u_{p_k,n}$ there exists a subsubsequence still denoted with $u_{p_k,n}$ such that, as $k \longrightarrow \infty$, $u_{p_k,n} \longrightarrow u_{\infty,n}$ weakly in $W^{1,r}(\Omega_\alpha^n)$, being $u_{\infty,n}$ a maximizer of Problem $(\mathcal{P}_n)$.*

In both previous theorems we have required that the final convex sets (i.e. the sets in which we search the optimum of the corresponding limit problems) were non-empty. We will prove that assuming $\mathcal{K}^\infty \neq \emptyset$ we can construct a sequence $\varphi_n \to \varphi$ in $C(\overline{\Omega}_\alpha)$ such that the corresponding set $\mathcal{K}_n^\infty$ is non-empty.

More precisely, the following result holds.

**Proposition 3.2** *Assume that $\mathcal{K}^\infty$ defined in (3.4) is non-empty. Then there exists a sequence of functions $\varphi_n \in C(\overline{\Omega}_\alpha)$ converging to $\varphi$ in $C(\overline{\Omega}_\alpha)$ such that $\mathcal{K}_n^\infty \neq \emptyset$, where $\mathcal{K}_n^\infty$ is the convex defined in (3.10).*

***Proof*** The proof in divided into two steps.

**Step 1** Let us show that the distance function $d(x) := dist(x, \partial\Omega_\alpha)$, $\forall x \in \overline{\Omega}_\alpha$ is an element of $\mathcal{K}^\infty$.

First of all we have that $d(x) = |x - x^*|$, for some $x^* \in \partial\Omega_\alpha$, is a 1-Lipschitz function in $\overline{\Omega}_\alpha$. In fact, considering $x, y \in \overline{\Omega}_\alpha$ and, for instance, $d(x) > d(y)$, we have

$$\begin{aligned} |d(x) - d(y)| = d(x) - d(y) = |x - x^*| - |y - y^*| \leq |x - y^*| - |y - y^*| \\ \leq |x - y|, \ \text{with } x^*, y^* \in \partial\Omega_\alpha. \end{aligned}$$

Moreover $d(\bar{x}) = 0, \forall \bar{x} \in \partial\Omega_\alpha$, for the definition of distance.

By assumption we have that $\mathcal{K}^\infty \neq \emptyset$, i.e. $\exists w \in \mathcal{K}^\infty$.

We have to show only that $\varphi(x) \leq d(x), \forall x \in \Omega_\alpha$.

Let us consider $x \in \overline{\Omega}_\alpha$, then

$$\varphi(x) = \varphi(x) - 0 \leq w(x) - w(y) \leq |x - y|, \forall y \in \partial\Omega_\alpha.$$

In particular:

$$\varphi(x) \leq \min_{y \in \partial\Omega_\alpha} |x - y| = d(x), \ \forall x \in \overline{\Omega}_\alpha. \tag{3.11}$$

**Step 2** Let us consider the function

$$d_n(x) := \begin{cases} dist(x, \partial\Omega_\alpha^n), & x \in \overline{\Omega}_\alpha^n \\ 0, & x \in \overline{\Omega}_\alpha \setminus \Omega_\alpha^n \end{cases}.$$

Analogously to what we saw in Step 1, we have that $d_n(x)$ is a 1-Lipschitz function in $\overline{\Omega}_\alpha$ (thus in $\overline{\Omega}_\alpha^n$) and $d_n(\bar{x}) = 0, \forall \bar{x} \in \partial\Omega_\alpha^n$.

To conclude our proof we construct a sequence of functions $\varphi_n(x) \in C(\overline{\Omega}_\alpha)$ such that:

$$\begin{aligned} &\text{(a) } d_n(x) \geq \varphi_n(x), \ \forall x \in \overline{\Omega}_\alpha; \\ &\text{(b) } \varphi_n \to \varphi \text{ in } C(\overline{\Omega}_\alpha). \end{aligned} \tag{3.12}$$

Let us define,

$$\varphi_n(x) = \varphi(x) - a_n(x) \tag{3.13}$$

with

$$a_n(x) = d(x) - d_n(x). \tag{3.14}$$

Let us prove (a).

(a) Let $x \in \overline{\Omega}_\alpha$ from (3.13) and (3.14), we have

$$\varphi_n(x) = \varphi(x) - d(x) + d_n(x) \leq d_n(x),$$

since $\varphi(x) \leq d(x), \ \forall x \in \overline{\Omega}_\alpha$.

To prove (b) we will show that $a_n(x) \to 0$ in $C(\overline{\Omega}_\alpha)$.

(b) First, given $x \in \overline{\Omega}_\alpha^n$, let us denote with $x_n \in \partial\Omega_\alpha^n$ and $x^* \in \partial\Omega_\alpha$ the points such that

$$d_n(x) = |x - x_n| \text{ and } d(x) = |x - x^*|.$$

Now, since $\Omega_\alpha^n \subset \Omega_\alpha$, $\forall n \in \mathbb{N}$, then $dist(x, \partial\Omega_\alpha^n) = dist(x, \overline{\Omega}_\alpha \setminus \Omega_\alpha^n)$, so we have

$$|x - x_n| \leq |x - y|, \forall y \in \overline{\Omega}_\alpha \setminus \Omega_\alpha^n.$$

Then, by choosing $y = x^*$, we obtain that $a_n(x) \geq 0$ (actually it holds $\forall x \in \overline{\Omega}_\alpha$).

So, we prove that

$$\max_{\overline{\Omega}_\alpha} |d(x) - d_n(x)| \to 0, \text{ as } n \to \infty$$

(we note that $\forall x \in \Omega_\alpha \; \exists n \in \mathbb{N} : x \in \Omega_\alpha^n$).

Since $x_n \in K^{i|n}$, for some $i|n$, let us consider $y \in \partial\Omega_\alpha^n \cap \partial\Omega_\alpha \cap \overline{\Sigma}_{j,+}^{i|n}$ (for some $j \in \{1, 2, 3\}$), where $K^{i|n}$ and $\Sigma_{j,+}^{i|n}$ are defined in Sect. 2. Then we have

$$|d(x) - d_n(x)| = d(x) - d_n(x) \leq |y - x| - |x - x_n| \leq |y - x_n| \leq diam(\Sigma_{j,+}^{i|n})$$
$$\leq \frac{1}{3^n} \to 0, n \to \infty$$

and the proof is concluded.

□

*Remark 3.3* From the step 1 of the proof of Proposition 3.2 we deduce that

$$w(x) \leq d(x), \;\forall w \in \mathcal{K}^\infty.$$

**Proposition 3.3** *Let us assume $\mathcal{K}^\infty \neq \emptyset$. Then there exists $u_{\infty,n}$ maximizer of Problem $(\mathcal{P}_n)$.*

***Proof*** By Proposition 3.2 we known that $\varphi_n(x)$ defined in (3.13) converges to $\varphi$ in $C(\overline{\Omega}_\alpha)$ and $\mathcal{K}_n^\infty$, defined in (3.10), is non-empty. Then we deduce that $\mathcal{K}_{p,n} \neq \emptyset$; moreover, by Theorem 3.2, we have that a maximizer $u_{\infty,n}$ of Problem $(\mathcal{P}_n)$ exists. □

Until now, we have studied the asymptotic behavior for $p \to \infty$, both in fractal and in pre-fractal case. The following theorem states the asymptotic behavior of the solutions in the case of $n \to \infty$ and $p = \infty$.

Now, let us define

$$\tilde{u}_{\infty,n}(x) = \begin{cases} u_{\infty,n}, & x \in \Omega_\alpha^n \\ 0, & x \in \Omega_\alpha \setminus \Omega_\alpha^n, \end{cases} \tag{3.15}$$

where $u_{\infty,n}$ is a maximizer of Problem $(\mathcal{P}_n)$.

**Theorem 3.3** *Let* $f \in L^1(\Omega_\alpha)$, $\mathcal{K}^\infty \neq \emptyset$, $\varphi \in C(\overline{\Omega}_\alpha)$, $\varphi_n(x) = \varphi(x) - a_n(x)$ *(with* $a_n(x)$ *defined in (3.14)) and* $\varphi(x) = d(x)$, $\forall x \in \partial\Omega_\alpha^n$ *and* $n \in \mathbb{N}$. *Then, any subsequence of functions* $\tilde{u}_{\infty,n}$ *defined in (3.15) admits a subsubsequence* $\star$*-weakly converging, as* $n \to \infty$, *in* $W^{1,\infty}(\Omega_\alpha)$ *to a maximizer* $u_\infty$ *of Problem* $(\mathcal{P})$.

***Proof*** Let $u_{\infty,n}$ be a maximizer of Problem $(\mathcal{P}_n)$. We note that $\tilde{u}_{\infty,n}$ defined in (3.15) is bounded in $W^{1,\infty}(\Omega_\alpha)$. Then, we deduce that any subsequence of $\tilde{u}_{\infty,n}$ admits a subsubsequence, again denoted with $\tilde{u}_{\infty,n}$, $\star$-weakly converging in $W^{1,\infty}(\Omega_\alpha)$ to a function $\tilde{v} \in W^{1,\infty}(\Omega_\alpha)$. Then

$$||\nabla\tilde{v}||_{L^\infty(\Omega_\alpha)} \leq 1.$$

For any $w \in \mathcal{K}^\infty$, now, we construct $w_n \in \mathcal{K}_n^\infty$ such that

$$\lim_{n\to\infty} \int_{\Omega_\alpha^n} f w_n \, \mathrm{d}x = \int_{\Omega_\alpha} f w \, \mathrm{d}x.$$

First, we observe that (see Proposition 3.2) the sequence $\varphi_n = \varphi - a_n$ converge to $\varphi$ in $C(\overline{\Omega}_\alpha)$, where $a_n(x) = d(x) - d_n(x)$, $x \in \overline{\Omega}_\alpha$.
By (3.11) and (a) in (3.12) and we have that

$$\varphi_n(x) \leq d_n(x), \forall x \in \overline{\Omega}_\alpha.$$

Then, $\forall x \in \overline{\Omega}_\alpha$, we define

$$w_n(x) := w(x) \wedge d_n(x) = \begin{cases} w(x), & \text{if } w(x) \leq d_n(x) \\ d_n(x), & \text{if } w(x) > d_n(x) \end{cases}.$$

Let us show that $w_n \in \mathcal{K}_n^\infty$.

By the fact that $w$ and $d_n$ are 1-Lipschitz function, we have that $w_n$ 1-Lipschitz function too. Since $w \geq \varphi \geq \varphi_n$, we deduce that $w \geq \varphi_n$ and then $w_n \geq \varphi_n$. Finally, from the assumption $\varphi(x) = d(x)$, $\forall x \in \partial\Omega_\alpha^n$ and $n \in \mathbb{N}$, we get $\varphi_n(x) = 0$ on $\partial\Omega_\alpha^n$ and then we obtain that $w_n \in \mathcal{K}_n^\infty$.

Moreover, as $d_n \to d$ in $C(\overline{\Omega}_\alpha)$, we obtain that $w_n \to w \wedge d$ in $C(\overline{\Omega}_\alpha)$, where $w \wedge d = w$ for Remark 3.3. In particular,

$$\lim_{n\to\infty} \int_{\Omega_\alpha^n} f w_n \, \mathrm{d}x = \int_{\Omega_\alpha} f w \, \mathrm{d}x.$$

Then we conclude how in the proof of Theorem 4.2 in [2]. □

Finally, we summarize all the previous asymptotic results in the scheme in Fig. 4.

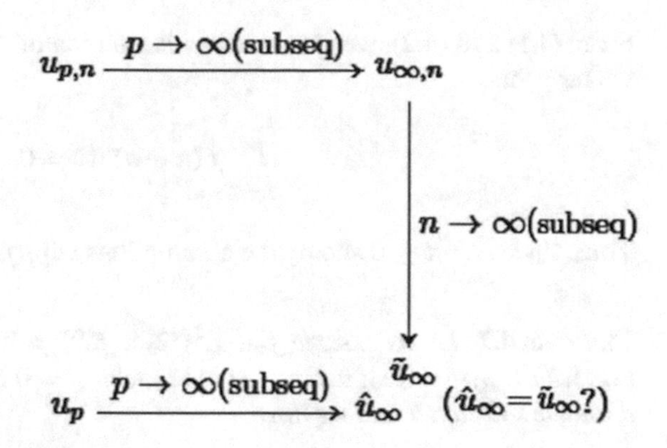


**Fig. 4** Summary of asymptotics

## 4 Uniqueness and Perspectives

The results of Sect. 3 show us that a solution of Problem $(\mathcal{P})$ can be obtained performing first the limit for $p \to \infty$ and then the limit for $n \to \infty$ of the solutions of Problems $(\mathcal{P}_{p,n})$, see Theorems 3.2 and 3.3. Moreover, we can obtain a solution of Problem $(\mathcal{P})$ also as limit for $p \to \infty$ of the solutions of Problems $(\mathcal{P}_p)$, see Theorem 3.1.

The functions $\tilde{u}_\infty$ and $\hat{u}_\infty$ solve the same problem but, in general, since we do not have uniqueness results, we cannot affirm that they are equal (see Fig. 4).

In [4] the authors give sufficient conditions that allows to obtain the convergence, as $p \to \infty$, of the whole sequence $u_{p,n}$ to $u_{\infty,n}$, solutions of Problems $(\mathcal{P}_n)$, and of the whole sequence $u_p$ to $u_\infty$, solution of the Problem $(\mathcal{P})$.

In the following theorems we give sufficient conditions which guarantee that Problem $(\mathcal{P})$ has only one solution. We note that we do not required smoothness for the datum and the obstacle. Moreover, we stress the fact that in the following results we obtain an explicit form for the solution.

We refer to [1] and [11] for the case of the equation.

**Theorem 4.1** *Let us assume $f \in L^1(\Omega_\alpha)$, $\mathcal{K}^\infty \neq \emptyset$ and $\varphi \in C(\overline{\Omega}_\alpha)$. If $f \geq 0$ in $\Omega_\alpha$, with $supp f = \overline{\Omega}_\alpha$, then Problem $(\mathcal{P})$ has a unique solution u and we have*

$$u(x) = d(x, \partial\Omega_\alpha), \;\; x \in \Omega_\alpha.$$

***Proof*** Let $u$ a solution of Problem $(\mathcal{P})$. Since $d(x, \partial\Omega_\alpha) := d(x) \in \mathcal{K}^\infty$, we have

$$\int_{\Omega_\alpha} f u \,\mathrm{d}x \geq \int_{\Omega_\alpha} f d \,\mathrm{d}x. \tag{4.1}$$

Moreover, since $u(x) \leq d(x)$ $\forall x \in \Omega_\alpha$ (see Remark 3.3), by the assumption on the sign of $f$ and from the monotonicity property of the integral, we deduce

$$\int_{\Omega_\alpha} f u \,\mathrm{d}x \leq \int_{\Omega_\alpha} f d \,\mathrm{d}x. \tag{4.2}$$

From (4.1) and (4.2), we obtain that $d$ is solution of the problem $(\mathcal{P})$. In particular, we have that

$$\int_{\Omega_\alpha} f(u-d)\,\mathrm{d}x = 0.$$

Thus, since $u - d \leq 0$ and by the assumptions $suppf = \overline{\Omega}_\alpha$, it follows that $u = d$. □

**Theorem 4.2** *Let us assume $f \in L^1(\Omega_\alpha)$, $\mathcal{K}^\infty \neq \emptyset$ and $\varphi \in C(\overline{\Omega}_\alpha)$. If $f \leq 0$ in $\Omega_\alpha$, with $suppf = \overline{\Omega}_\alpha$, and $\varphi \leq 0$ in $\Omega_\alpha$ with $\varphi = 0$ in $\partial\Omega_\alpha$, then Problem $(\mathcal{P})$ has a unique solution $u$ and we have*

$$u(x) = \max_{y\in\overline{\Omega}_\alpha}\{\varphi(y) - |x-y|\}, \ \ x \in \Omega_\alpha.$$

***Proof*** Let

$$\overline{\varphi}(x) = \max_{y\in\overline{\Omega}_\alpha}\{\varphi(y) - |x-y|\}, x \in \overline{\Omega}_\alpha.$$

and $\mathcal{K}^\infty$ the set defined in relation (3.4). We have that $\overline{\varphi} \in \mathcal{K}^\infty$, since $\overline{\varphi} \in Lip_1(\Omega_\alpha)$, $\overline{\varphi}(x) \geq \varphi(x), \forall x \in \overline{\Omega}_\alpha$ and $\overline{\varphi}(\bar{x}) = 0, \forall \bar{x} \in \partial\Omega_\alpha$ (it follows from the sign of $\varphi$ and choosing $y = \bar{x}$). Now, let $u$ a solution of Problem $(\mathcal{P})$. Since $\overline{\varphi} \in \mathcal{K}^\infty$, then

$$\int_{\Omega_\alpha} fu\,\mathrm{d}x \geq \int_{\Omega_\alpha} f\overline{\varphi}\,\mathrm{d}x. \tag{4.3}$$

Nevertheless, we have $u \geq \overline{\varphi}$ in $\Omega_\alpha$. Then from the assumption on the sign of $f$ and from the monotonicity property of the integral, we deduce

$$\int_{\Omega_\alpha} fu\,\mathrm{d}x \leq \int_{\Omega_\alpha} f\overline{\varphi}\,\mathrm{d}x. \tag{4.4}$$

From (4.3) and (4.4), we obtain that $\overline{\varphi}$ is solution of the problem $(\mathcal{P})$. In particular, we have that

$$\int_{\Omega_\alpha} f(u-\overline{\varphi})\,\mathrm{d}x = 0.$$

Thus, since $u - \overline{\varphi} \geq 0$ and from the assumptions on $f$, it follows that $u = \overline{\varphi}$. In fact, let us assume by contradiction that

$$\exists x^* \in \Omega_\alpha \text{ such that } u(x^*) - \overline{\varphi}(x^*) > 0,$$

for the theorem of the permanence of the sign there exists $U = U(x^*)$ such that

$$u(x) - \overline{\varphi}(x) > 0, \forall x \in U$$

then

$$0 = \int_{\Omega_\alpha} f(u - \overline{\varphi})\,\mathrm{d}x \leq \int_U f(u - \overline{\varphi})\,\mathrm{d}x \leq m \int_U f\,\mathrm{d}x \leq 0,$$

with $m = \inf_U (u - \overline{\varphi})$. □

*Remark 4.1* The previous Theorems 4.1 and 4.2 show uniqueness results in weaker assumption about regularity of $f$ and $\varphi$ with respect to the ones required in [4]. We note that the proof uses the assumption on the sign of $f$. In particular we need that $f \leq 0$ (or $f \geq 0$) and that it is not equal to 0 on a set with positive measure.

The study of the cases in which $f$ chances sign in $\Omega_\alpha$ or can be equal to 0 in a set non-having measure equal to 0 is still open.

In the same way, we can prove the following analogous results of Theorems 4.1 and 4.2 in the case of pre-fractal approximating domain $\Omega_\alpha^n$.

**Theorem 4.3** *Let us assume* $f \in L^1(\Omega_\alpha)$, $\mathcal{K}^\infty \neq \emptyset$ *and* $\varphi \in C(\overline{\Omega}_\alpha^n)$. *If* $f \geq 0$ *in* $\Omega_\alpha^n$, *with* $supp f = \overline{\Omega}_\alpha^n$, *then Problem* $(\mathcal{P}_n)$ *has a unique solution* $u_{\infty,n}$ *and we have*

$$u_{\infty,n}(x) = d(x, \partial\Omega_\alpha^n), \ x \in \Omega_\alpha^n.$$

**Theorem 4.4** *Let us assume* $f \in L^1(\Omega_\alpha)$, $\mathcal{K}_n^\infty \neq \emptyset$ *and* $\varphi_n \in C(\overline{\Omega}_\alpha^n)$. *If* $f \leq 0$ *in* $\Omega_\alpha^n$, *with* $supp f = \overline{\Omega}_\alpha^n$, $\varphi_n \leq 0$ *in* $\Omega_\alpha^n$ *and* $\varphi_n = 0$ *in* $\partial\Omega_\alpha^n$, *then Problem* $(\mathcal{P}_n)$ *has a unique solution* $u_{\infty,n}$ *and we have*

$$u_{\infty,n}(x) = \max_{y \in \overline{\Omega}_\alpha^n} \{\varphi_n(y) - |x - y|\}, x \in \Omega_\alpha^n.$$

*Remark 4.2* We note that in Theorem 4.4 we require that $\mathcal{K}_n^\infty \neq \emptyset$ in place of $\mathcal{K}^\infty \neq \emptyset$. This because, with our definition of $\varphi_n$ (see (3.13)) the assumption $\varphi_n = 0$ on $\partial\Omega_\alpha^n$ cannot be satisfied if $\varphi < 0$ in $\Omega_\alpha$.

Thanks these uniqueness results, we obtain that starting to $u_{p,n}$ or starting to $u_p$ and performing the limits, as $p \to \infty$ and $n \to \infty$, we obtain the same solution of Problem $(\mathcal{P})$, see Fig. 5.

The aim of the next section is to state uniform estimates for the numerical approximating problem and discuss about the rate of vanishing of the approximation error. In some sense, the possibility to perform numerical analysis provides a further motivation to consider the corresponding problems in pre-fractal approximating domains.

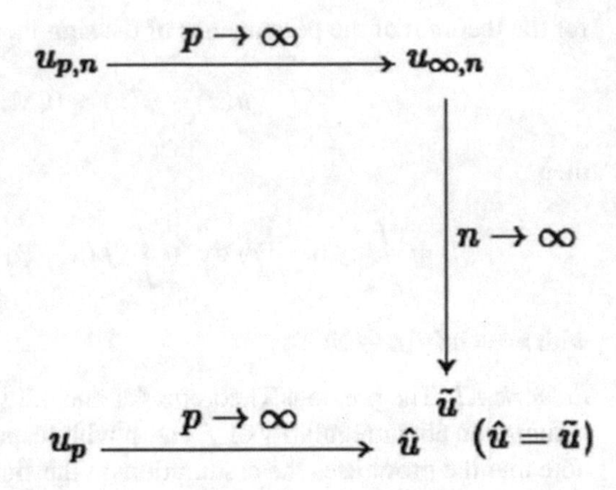


**Fig. 5** New summary of the results

# 5 Uniform and Error Estimates

In this section we will consider $n \in \mathbb{N}$ fixed, so we refer to $\Omega^n_\alpha$ writing simply $\Omega$. In order to define the approximate solution $u^h_{p,n}$ (that we will denote briefly with $u_h$) of the problem $(\mathcal{P}_{p,n})$ according to the Galerkin method, we introduce a triangulation $T_h$ of the domain $\Omega$. So, let us consider $T_h$ a partition of the domain $\Omega$ into disjoint, open and regular triangle $\tau$, each side bounded by $h$ and such that $\overline{\Omega} = \bigcup_{\tau \in T_h} \overline{\tau}$. Moreover, we require that $T_h$ is regular and conformal, i.e. that are satisfied the conditions of the following

**Definition 5.1** A triangulation $T_h$ of $\Omega$ is regular and conformal if

(1) $\bar{\Omega} = \bigcup\limits_{\tau \in T_h} \bar{\tau}$
(2) $\tau \neq \emptyset, \forall\ \tau \in T_h$
(3) $\tau_1 \cap \tau_2 = \emptyset, \quad \forall\ \tau_1, \tau_2 \in T_h : \quad \tau_1 \neq \tau_2$
(4) $\bar{\tau_1} \cap \bar{\tau_2} \neq \emptyset, \quad \tau_1 \neq \tau_2 \quad \Rightarrow \quad \tau_1 \cap \tau_2 =$ edge or vertex
(5) $\exists\, \sigma > 0$ such that $\max_{\tau \in T_h} \frac{h_\tau}{\eta_\tau} \leq \sigma$
where $h_\tau = diam(\tau)$ and $\eta_\tau = \sup\{diam(B) : B \text{ ball} \subset \tau\}$.

We consider the following finite dimensional space associated with $T_h$:

$$S_h = \left\{ v \in C(\bar{\Omega}) \,:\, v|_\tau \text{ is affine } \forall \tau \in T_h \right\}$$

and

$$S_{h,0} = \left\{ v \in S_h \,:\, v = 0 \text{ on } \partial\Omega \right\}.$$

Furthermore, with $\pi_h$ we denote the interpolation operator, $\pi_h : C(\bar{\Omega}) \to S_h$, such that $\pi_h v(V_i) = v(V_i)$ for any vertex $V_i$ of the partitioning $T_h$.

Now, let us consider the obstacle problem in the finite dimensional space $S_{h,0}$ :

$$\text{find } u_h \in \mathcal{K}_h \ : \int_\Omega |\nabla u_h|^{p-2} \nabla u_h \nabla (v_h - u_h) \mathrm{d}x - \int_\Omega f(v_h - u_h) \mathrm{d}x \geqslant 0, \forall v_h \in \mathcal{K}_h, \tag{$\mathcal{P}_h$}$$

where

$$\mathcal{K}_h = \{v_h \in S_{h,0} : v_h \geqslant \varphi_h \text{ in } \Omega\},$$

with $\varphi_h = \pi_h \varphi_n$. Before to state uniform estimates, we prove the following

**Proposition 5.1** *Let $\mathcal{K}_n^\infty$ be non-empty, where $\mathcal{K}_n^\infty$ is the set defined in (3.10), then*

$$\mathcal{K}_{n,h}^\infty = \{v_h \in S_{h,0} : v_h \geqslant \varphi_h \text{ in } \Omega, ||\nabla v_h||_\infty \leq 1\} \neq \emptyset.$$

***Proof*** The proof is divided into two part.

**Step 1** First, let us prove the thesis in the case of $\Omega = [a, b] \subset \mathbb{R}$.

Let $T_h = \{a = a_0 < a_1 < \ldots < a_i < \ldots < a_{n-1} < a_n = b\}$ be a partition of $[a, b]$, $w \in \mathcal{K}_n^\infty$ and $w_h = \pi_h w$. We have

(i) $w_h \in S_{h,0}$;
(ii) Let $I_j = [a_{j-1}, a_j]$, $j \in \{1, \ldots, n\}$ be a subinterval of $[a, b]$, then

$$w_h^j(x) = w_h(x)|_{I_j} = \pi_h w(x)|_{I_j} = \frac{a_j - x}{a_j - a_{j-1}} w(a_{j-1}) + \frac{x - a_{j-1}}{a_j - a_{j-1}} w(a_j) \geq$$

$$\geq \frac{a_j - x}{a_j - a_{j-1}} \varphi(a_{j-1}) + \frac{x - a_{j-1}}{a_j - a_{j-1}} \varphi(a_j) = \pi_h \varphi(x)|_{I_j}$$

$$= \varphi_h(x)|_{I_j} = \varphi_h^j(x), \ \forall x \in I_j, \ \forall j \in \{1, \ldots, n\}.$$

Hence, we have that $w_h \geq \varphi_h$.
(iii) Finally, let as show that Lipschitz condition $|w_h(x) - w_h(y)| \leq |x - y|$, $\forall x, y \in [a, b]$ hold. We have two possibilities:

(a) $x, y \in I_j$, $j \in \{1, \ldots, n\}$, so $\exists \lambda, \sigma \in [0, 1]$ such that $x = a_{j-1} + \lambda(a_j - a_{j-1})$ and $y = a_{j-1} + \sigma(a_j - a_{j-1})$. Since $w_h^j(x)| = w(a_{j-1}) + \frac{w(a_j) - w(a_{j-1})}{a_j - a_{j-1}}(x - a_{j-1})$, we have

$$|w_h(x) \ - w_h(y)| = |w_h^j(x) - w_h^j(y)| = |\lambda - \sigma||w(a_j)$$
$$-w(a_{j-1})| \leq |\lambda - \sigma||a_j - a_{j-1}| = |x - y|.$$

(b) $x \in I_j$ and $y \in I_k$, with $j, k \in \{1, \ldots, n\}$ and $j \neq k$. We can suppose $j < k$ (i.e. $x < y$), then $k = j + p$, $p \in \{1, \ldots, n - 1\}$; in particular

$\exists\lambda, \sigma \in [0,1]$ such that $x = a_{j-1} + \lambda(a_j - a_{j-1})$ and $y = a_{j+p-1} + \sigma(a_{j+p} - a_{j+p-1})$. Since

$$w_h^i(a_i) = w_h^{i+1}(a_i),\ i \in \{1, \dots, n-1\},$$

we get

$$|w_h(x) - w_h(y)| = |w_h^j(x) - w_h^{j+p}(y)| \le$$

$$\le |w_h^j(x) - w_h^j(a_j)| + |w_h^{j+1}(a_j) - w_h^{j+1}(a_{j+1})| + \ldots + |w_h^{j+p}(a_{j+p-1}) - w_h^{j+p}(y)|.$$

For the case (a), we obtain

$$|w_h(x) - w_h(y)| \le |x - a_j| + |a_j - a_{j+1}| + \ldots + |a_{j+p-1} - y| = |x - y|.$$

**Step 2** Now, let us consider $\Omega \subset \mathbb{R}^2$ one of the pre-fractal domains introduced in Sect. 2, $T_h$ a regular and conformal triangulation of $\Omega$ and $w \in \mathcal{K}_n^\infty$. Again we want to show that Lipschitz condition $|w_h(x) - w_h(y)| \le ||x - y||$, $\forall x, y \in \Omega$ holds.

(a) if $x, y \in \tau$, $\tau \in T_h$, since $T_h$ is convex, then the condition holds thanks to Step 1.
(b) if $x \in \tau_1$ and $y \in \tau_2$, $\tau_1, \tau_2 \in T_h$, $\tau_1 \neq \tau_2$, are not connected by a line segment, we can apply again Step 1 considering $R = [a,b] \times [c,d]$ such that $\Omega \subset R$ and the 1-Lipschitz function

$$w^*(s) = \max_{z \in R}\{w(z) - |s - z|\}$$

(see [17] for details and proofs).

□

**Theorem 5.1** *Let us assume $f \in L^1(\Omega_\alpha)$, $\mathcal{K}^\infty \neq \emptyset$ and let us consider $u_{p,n}^h$ solution of Problem $(\mathcal{P}_h)$. Then there exists a constant C, such that*

$$||u_{p,n}^h||_{1,p} \le C, \forall p > \bar{p} > 2,$$

*with C independent from p and h.*

***Proof*** Let us denote $u_{p,n}^h = u_h$ and consider $v_h \in \mathcal{K}_{n,h}^\infty$ (which is a feasible choice thanks to Propositions 3.2 and 5.1). By Remark 3.1, since $u_h$ is solution, retracing the same passages of the proof of Theorem 3.1, we obtain

$$||\nabla u_h||_p \le \Bigg\{2|\Omega| + 2p||f||_1||v_h||_\infty + \frac{(p-1)2^{\frac{p+1}{p-1}}||f||_1^{\frac{p}{p-1}}\pi^{-\frac{1}{p-1}}d^{\frac{p}{p-1}}}{p^{\frac{1}{p-1}}}$$

$$+(p-1)2^{\frac{p}{p-1}}||f||_1^{\frac{p}{p-1}}(\pi)^{-\frac{1}{p-1}}\left(\frac{p-1}{p-2}\right)\Bigg\}^{\frac{1}{p}} = C(p).$$

Let us consider $m > 2$ and let $C = \sup_{p>m}\{C(p)\}$ be, then

$$||\nabla u_h||_p \leq C.$$

Hence, by relation (3.3), we have

$$||u_h||_{1,p} \leq (d+1)||\nabla u_h||_p \leq C,$$

with $C$ independent from $p$ and $h$. □

*Remark 5.1* The search of sufficient conditions which guarantee the uniqueness, done in previous section, is itself interesting. Nevertheless, it is important in order to know an approximation of the solution of the final Problem $(\mathcal{P})$. In fact, as the FEM-solutions $u^h_{p,n}$ converge to solution $u_{p,n}$ for $h \to 0$ (see [3, 5] and the following Theorem 5.2 for the rate of convergence of the approximation error), they (in this framework it is important the uniqueness) are an approximation for the solution of Problem $(\mathcal{P})$ for which it is not possible to perform numerical analysis, being a problem in a fractal boundary domain.

Let us conclude this section discussing about the rate of convergence of the approximation error. In particular, we quote and report below the following proposition (see Theorem 5.1 in [3]), adapted to our case (that is one obstacle case). We recall that $n \in \mathbb{N}$ is fixed.

**Proposition 5.2** *Let us consider $p \in (2, \infty)$ and let $f \in L^{p'}(\Omega)$, with $\frac{1}{p} + \frac{1}{p'} = 1$. Let $u$ and $u_h$ be the solutions of Problems $(\mathcal{P}_{p,n})$ and $(\mathcal{P}_h)$ and let us assume*

$$u \in W^{s,2}(\Omega), \; 2 - \frac{1-2/p}{p-1} < s < 2, \tag{5.1}$$

$$\varphi_n \in W^{\sigma,p}(\Omega), \quad \sigma = s - 1 + 2/p \tag{5.2}$$

*and*

$$|\nabla u|^{-1} \in L^\infty(\Omega) \;\; if \;\; q = t \quad or \quad |\nabla u|^{-\frac{(p-t)q}{t-q}} \in L^1(\Omega) \;\; if \;\; q \in [1, t). \tag{5.3}$$

*Then*

$$||u - u_h||_{1,q} \leq ch^r(||u||^{2/t}_{s,2} + ||u||^{1/t}_{\sigma,p} + ||\varphi_n||^{1/t}_{\sigma,p}), \tag{5.4}$$

*where $r = \frac{\sigma}{t}$, $t \in [2, p]$, $q \in [1, t]$, and the constant $c$ does not depend on $h$.*

The assumptions of the previous proposition guarantee an optimal rate of vanishing for the approximation error.

Some improvements have been done in [5] following the approach of Grisvard (see [9]). In particular, in order to obtain a better rate of convergence, it need to

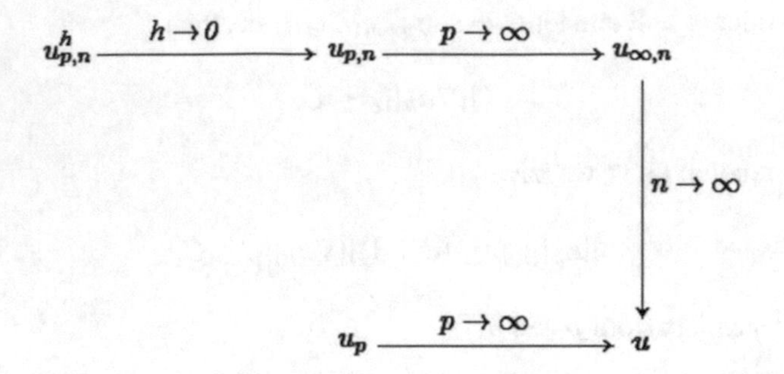


**Fig. 6** Final summary

make further assumption of the triangulation of the domain and on the regularity of the solution, which has to belong to suitable weighted Sobolev spaces.

Assuming satisfied all the assumptions of the theorems of previous sections, we can summarize all the results obtained in the following final scheme (see Fig. 6).

**Acknowledgments** The author thanks Prof Maria Agostina Vivaldi for fruitful and precious discussions.

## References

1. Bhattacharya, T., DiBenedetto, E., Manfredi, J.: Limits as $p \to +\infty$ of $\Delta_p u_p = f$ and related extremal problems. Some topics in nonlinear PDEs (Turin, 1989). Rend. Sem. Mat. Univ. Politec. Torino **47**, 15–68 (1989). Special Issue (1991)
2. Capitanelli, R., Fragapane, S.: Asymptotics for quasilinear obstacle problems in bad domains. Discrete Contin. Dyn. Syst. Ser. S **12**(1), 43–56 (2019)
3. Capitanelli, R., Vivaldi, M.A.: FEM for quasilinear obstacle problems in bad domains. ESAIM Math. Model. Numer. Anal. **51**, 2465–2485 (2017)
4. Capitanelli, R., Vivaldi, M.A.: Limit of p-Laplacian obstacle problems. Adv. Calc. Var. Published online: 15th April 2020. https://doi.org/10.1515/acv-2019-0058
5. Capitanelli, R., Fragapane, S., Vivaldi, M.A.: Regularity results for p-Laplacians in pre-fractal domains. Adv. Nonlinear Anal. **8**(1), 1043–1056 (2019)
6. Diaz, J.I.: Nonlinear Partial Differential Equations and Free Boundaries, vol. I. Elliptic Equations. Research Notes in Mathematics, vol. 106. Pitman, Boston (1985)
7. Evans, L.C.: Partial Differential Equations. Graduate Studies in Mathematics, vol. 19. American Mathematical Society, Providence (1998)
8. Evans, L.C., Gangbo, W.: Differential equations methods for the Monge-Kantorovich mass transfer problem. Mem. Am. Math. Soc. **137**(653), 1–66 (1999)
9. Grisvard, P.: Elliptic Problems in Nonsmooth Domains. Monographs and Studies in Mathematics, vol. 24. Pitman, Boston (1985)
10. Hutchinson, J.E.: Fractals and selfsimilarity. Indiana Univ. Math. J. **30**(5), 713–747 (1981)
11. Ishii, H., Loreti, P.: Limits of solutions of p-Laplace equations as p goes to infinity and related variational problems. SIAM J. Math. Anal. **37**(2), 411–437 (2005)
12. Kawohl, B.: On a family of torsional creep problems. J. Reine Angew. Math. **410**, 1–22 (1990)

13. Leoni, G.: A First Course in Sobolev Spaces. Graduate Studies in Mathematics, vol. 105. American Mathematical Society, Providence (2009)
14. Mandelbrot, B.B.: The Fractal Geometry of Nature. W. H. Freeman, San Francisco (1982)
15. Mandelbrot, B.B.: Fractals and Scaling in Finance. Springer, Berlin (1997)
16. Mazón, J.M., Rossi, J.D., Toledo, J.: Mass transport problems for the Euclidean distance obtained as limits of p-Laplacian type problems with obstacles. J. Differ. Equ. **256**, 3208–3244 (2014)
17. McShane, E.J.: Extension of range of functions. Bull. Am. Math. Soc. **40**(12), 837–842 (1934)
18. Mosco, U., Vivaldi, M.A.: Layered fractal fibers and potentials. J. Math. Pures Appl. (9) **103**(5), 1198–1227 (2015)
19. Royden, H.L.: Real Analysis, 2nd edn. Macmillan, New York (1968)
20. Troianiello, G.M.: Elliptic Differential Equations and Obstacle Problems. The University Series in Mathematics. Plenum Press, New York (1987)

# Discretization of the Koch Snowflake Domain with Boundary and Interior Energies

**Malcolm Gabbard, Carlos Lima, Gamal Mograby, Luke Rogers, and Alexander Teplyaev**

**Abstract** We study the discretization of a Dirichlet form on the Koch snowflake domain and its boundary with the property that both the interior and the boundary can support positive energy. We compute eigenvalues and eigenfunctions, and demonstrate the localization of high energy eigenfunctions on the boundary via a modification of an argument of Filoche and Mayboroda. Hölder continuity and uniform approximation of eigenfunctions are also discussed.

**Keywords** Koch snowflake domain · Laplacian · Landscape function · Localization · Discrete approximations

## 1 Introduction

The main objective of this paper is to investigate a discrete version of the eigenvalue problem $\Delta u = \lambda u$ on the Koch snowflake domain, which we denote by $\Omega$. To this end, we follow [30] and introduce a Dirichlet form (with a suitable domain) on $\Omega$,

$$\mathscr{E}(u) := \int_{\Omega} (\nabla u)^2 d\mathcal{L}^2 + \mathscr{E}_{\partial\Omega}(u|_{\partial\Omega}),$$

M. Gabbard
Colorado College, Department of Mathematics and Computer Science, Colorado Springs, CO, USA
e-mail: malcolm.gabbard@coloradocollege.edu

C. Lima
California State University Long Beach, Department of Physics and Astronomy, Long Beach, CA, USA
e-mail: carlos.lima@student.csulb.edu

G. Mograby · L. Rogers · A. Teplyaev (✉)
University of Connecticut, Department of Mathematics, Storrs, CT, USA
e-mail: gamal.mograby@uconn.edu; luke.rogers@uconn.edu; alexander.teplyaev@uconn.edu



M. R. Lancia, A. Rozanova-Pierrat (eds.), *Fractals in Engineering: Theoretical Aspects and Numerical Approximations*, SEMA SIMAI Springer Series 8,
https://doi.org/10.1007/978-3-030-61803-2_4

where $\mathcal{L}^2$ is the usual Lebesgue measure on $\mathbb{R}^2$ and $\mathcal{E}_{\partial\Omega}$ denotes the Kusuoka–Kigami Dirichlet form on the Koch snowflake boundary $\partial\Omega$. The novelty of this approach by comparison with past work [22, 26, 40, 44, 45, 57] is that both the Euclidean interior and the fractal boundary carry non-trivial Dirichlet forms. We approximate the Dirichlet form $\mathcal{E}$ by a sequence of discrete energies $\mathcal{E}_n$ (Theorem 4.1). This is done by inductively constructing triangulations of $\overline{\Omega} = \Omega \cup \partial\Omega$, the edges of which are then treated as a sequence of finite planar graphs $\Gamma_n$ equipped with a discrete energy $\mathcal{E}_n$ and measure $m_n$, and hence an inner product $\langle \cdot, \cdot \rangle$. We then define a discrete Laplacian $L_n$ such that

$$\mathcal{E}_n(u, v) = -\langle L_n u, v \rangle_n . \tag{1.1}$$

Note that $L_n$ takes into account both the interior $\Omega$ and fractal boundary $\partial\Omega$. Our approach is related to several recent works on diffusion problems involving fractal membranes [15, 36, 37]. Moreover, it can be viewed as a generalization of work of Lapidus et. al. [45] on the eigenstructure of the Dirichlet Laplacian on $\Omega$. Indeed, our numerical results on the spectra of $L_n$ are closely related to those for a discretization of the eigenvalue problem $\Delta u = \lambda u$ with Dirichlet boundary conditions due to a localization phenomenon that will be explored in Sect. 5.

One of the main goals of this paper is to investigate the impact of the fractal boundary on the eigenmodes of $L_n$. This problem is of general interest, particularly in physics where questions like "How do ocean waves depend on the topography of the coastlines?" and "How do trees and wind interact?" have already been studied. For these and more examples the reader is referred to [54, 55] and references therein. We note that Sapoval [53] conducted an experimental investigation of acoustic vibration modes of with soap bubbles placed on fractal drums, and made the striking observation that the fractal boundary causes some low-frequency wave modes to localize. Placing a soap bubble on the fractal boundary of a drum is mathematically equivalent to imposing Dirichlet boundary conditions; our situation is somewhat different, in that our model allows for non-trivial boundary energy, but nevertheless we find significant localization effects, which may be summarized as follows:

- The eigenvalue counting function has two regimes with different scaling. The threshold for the change in regimes is approximately the largest eigenvalue of the Dirichlet Laplacian on $\Omega$.
- Eigenfunctions with eigenvalues of $L_n$ below the threshold are not localized.
- Eigenfunctions with eigenvalues of $L_n$ above the threshold begin to show localization near $\partial\Omega$. As the eigenvalue increases, so does this localization.
- Eigenfunctions corresponding to eigenvalues of $L_n$ that are significantly above the threshold are strongly localized on $\partial\Omega$.

A class of boundary-localized high frequency eigenmodes known as *whispering-gallery modes* are well understood for convex domains, though an analytic explanation for their appearance in more general domains is not yet known [26]. However, our boundary modes do not seem to be whispering-gallery modes, or any other known class of localized high frequency modes. Rather, they appear to arise because the boundary part of the Dirichlet form and the interior part have different scalings and hence interact only weakly. An elementary way to see that such modes should appear in our model is given using a variant of the landscape map of Filoche and Mayboroda [10, 20] in Sect. 6.

The results given here are part of a long term study that aims to provide robust computational tools to address, in a fractal setting, a number of linear and nonlinear problems arising from physics [5–7, 12, 18, 25]. The physics of magnetic fields, and of vector equations more generally, are particularly challenging on fractal spaces. Moreover, a discretization of the type considered here is expected to be essential in studying quantum walks [3, 4, 11, 32, 51]. On the abstract mathematical side, our work is related to [1, 2, 8, 13, 14, 17, 23, 27–29, 46, 47, 52, 56] and to the classical Venttsel's problem [9, 58].

The paper is organized as follows. Section 2, follows the treatment in [30] to introduce a Dirichlet form on the Koch snowflake. In Sect. 3 we give the Dirichlet form on the snowflake domain and discuss some of its properties, such as the Hölder continuity and uniform approximation of eigenfunctions. In Sect. 4 we construct a triangular grid to approximate the Koch snowflake domain and introduce the discrete Laplacian $L_n$. Section 5 is concerned with our algorithm and numerical results, including the existence of boundary-localized eigenfunctions and their effect on the eigenvalue counting function. Finally, in Sect. 6 we show that the localized eigenfunctions $L_n$ can be predicted numerically using a variant of an argument from [20].

## 2 Dirichlet Form on the Koch Snowflake

The Koch snowflake and the associated snowflake domain are well-known. Here we introduce some notation and foundational results for our analysis, following [30]. Let $\{F_i\}_{i=1}^4$ be the iterated function system defined on $\mathbb{C}$ by

$$F_1(z) = \frac{z}{3} \qquad F_2(z) = \frac{z}{3}e^{i\frac{\pi}{3}} + \frac{1}{3}$$

$$F_3(z) = \frac{z}{3}e^{-i\frac{\pi}{3}} + \frac{3+i\sqrt{3}}{6} \qquad F_4(z) = \frac{z+2}{3}.$$

The Koch curve is the unique non-empty compact subset $K$ of $\mathbb{C}$ such that $K = \bigcup_{i=1}^4 F_i(K)$. It can be approximated by a sequence of finite graphs for which the following notation is convenient.

**Definition 2.1** Let $S = \{1, 2, 3, 4\}$. We define $W = S^{\mathbb{N}}$ and call $\omega \in W$ an infinite word. Similarly, a finite word of length $n \in \mathbb{N}$ is $w \in S^n$; we write $|w| = n$ for its length.

We write $F_w := F_{w_n} \circ \ldots \circ F_{w_1}$, where $w = w_1 \ldots w_n \in S^n$ and introduce finite graphs approximating the Koch curve as follows.

**Definition 2.2** Let $V_0(K) = \{0, 1\} \subset \mathbb{C}$ and $V_w(K) := F_w(V_0(K))$ for a finite word $w$. Then define

$$V_n(K) := \bigcup_{|w|=n} V_w(K), \text{ and } V_*(K) := \bigcup_{n \geq 0} V_n(K).$$

We consider the points of $V_n(K)$ as vertices of a graph in which adjacency, denoted by $p \sim_n q$, means that there is a word $w$ of length $n$ such that $p, q \in V_w(K)$.

On each of these graphs we define a graph energy for $u : V_*(K) \to \mathbb{R}$ by

$$\mathscr{E}_K^{(n)}(u) = \frac{4^n}{2} \sum_{p \in V_n(K)} \sum_{q \underset{n}{\sim} p} (u(q) - u(p))^2. \tag{2.1}$$

Following the general treatment in [34], to which we refer for all omitted details, we see that $\{\mathscr{E}_K^{(n)}(u)\}$ is nondecreasing, so $\mathscr{E}(u) = \lim_{m \to \infty} \mathscr{E}_m(u)$ is well defined; setting its domain to be $\{u : V_*(K) \to \mathbb{R} \mid \mathscr{E}_K(u) < \infty\}$ one obtains a resistance form. This non-negative definite, symmetric quadratic form extends to $\mathcal{D}(\mathscr{E}_K) := \{u \in \mathscr{C}(K) \mid \mathscr{E}_K(u|_{V_*}) < \infty\}$, where $\mathscr{C}(K)$ is the space of continuous functions on $K$. There is a resistance metric $R(x, y)$ on $K$ defined from $\mathscr{E}_K$ and with the property that points $x, y \in V_n$ with $x \sim_n y$ have $R(x, y)$ comparable to $4^{-n}$, and thus $R(x, y)$ is bi-Hölder to the Euclidean metric on $K$ with $R(x, y) \asymp |x - y|^{\frac{\log 4}{\log 3}}$. There is a resistance estimate for $f \in \mathcal{D}(\mathscr{E}_K)$

$$|f(x) - f(y)|^2 \leq R(x, y)\mathscr{E}_K(f) \tag{2.2}$$

so in particular these functions are $\frac{\log 2}{\log 3}$-Hölder in the Euclidean metric, see also [24, Corollary 4.8].

Continuing our use of results from [34], we see that if the Koch curve is equipped with the standard Bernoulli probability measure $\mu_K$, i.e. the self-similar measure with weights $\{\mu_i\}_{i=1}^4$ and $\mu_i = \frac{1}{4}$ for $i \in \{1, 2, 3, 4\}$, then $(\mathcal{D}(\mathscr{E}_K), \mathscr{E}_K)$ gives a strongly local regular Dirichlet form on $L^2(K, \mu_K)$.

A particular collection of functions in $\mathcal{D}(\mathscr{E}_K)$ of $\mathscr{E}_K$ will be useful in what follows. A function $h$ on $K$ is called harmonic if it minimizes the graph energies $\mathscr{E}_K^{(n)}(h)$ for all $n \geq 1$. It is called piecewise harmonic at scale $n$ if it is harmonic on the complement of $V_n$, or equivalently if $h \circ F_w^{-1}$ is harmonic for each word $w$ of length $n$. Piecewise harmonic functions are uniform-norm dense in $\mathscr{C}(K)$ and dense in $\mathcal{D}(\mathscr{E}_K)$ with respect to the norm $\left(\|u\|_{L^2(K,\mu_K)}^2 + \mathscr{E}_K(u)\right)^{1/2}$.

As in [24], we transfer the above definitions for the Koch curve to the boundary of the snowflake domain $\overline{\Omega}$ in the obvious manner. Write the boundary $\partial\Omega$ as a union $\cup_j K_j$ of three congruent copies of $K$. Specifically, let $K_j = \varphi_j(K)$ where

$\varphi_j$ is the Euclidean translation and rotation such that $0 \mapsto \sqrt{3}e^{i(4j-3)\pi/6}$ and $1 \mapsto \sqrt{3}e^{i(4j+1)\pi/6}$.

**Definition 2.3** The boundary energy and its domain are defined by

$$\mathcal{D}(\mathscr{E}_{\partial\Omega}) := \left\{ u : \partial\Omega \to \mathbb{R} \mid u|_{K_i} \circ \varphi_i^{-1} \in \mathcal{D}(\mathscr{E}_K), i = 1, \ldots, 3 \right\},$$

$$\mathscr{E}_{\partial\Omega}(u) := \mathscr{E}_K(u|_{K_1} \circ \varphi_1^{-1}) + \mathscr{E}_K(u|_{K_2} \circ \varphi_2^{-1}) + \mathscr{E}_K(u|_{K_3} \circ \varphi_3^{-1}) \text{ if } u \in \mathcal{D}(\mathscr{E}_{\partial\Omega}).$$

We then let

$$\mu(\cdot) := \mu_K(\varphi_1^{-1}(\cdot)) + \mu_K(\varphi_2^{-1}(\cdot)) + \mu_K(\varphi_3^{-1}(\cdot)) \tag{2.3}$$

so that $(\mathscr{E}_{\partial\Omega}, \mathcal{D}(\mathscr{E}_{\partial\Omega}))$ is a strongly local Dirichlet form on $L^2(\partial\Omega, \mu)$. Since $\mu_K(K) = 1$ we have $\mu(\partial\Omega) = 3$.

# 3 Dirichlet Form on the Snowflake Domain

We wish to consider a Dirichlet form on $\overline{\Omega}$ that incorporates our form $\mathscr{E}_{\partial\Omega}$ as well as the classical Dirichlet energy on $\Omega$. The latter is, of course, simply $\int_\Omega |\nabla f|^2 \, d\mathcal{L}^2$, where $\mathcal{L}^2$ is Lebesgue measure. The domain is the Sobolev space $H^1$. The fact that a nice Dirichlet form of this type exists depends on results of Wallin [59] and Lancia [36], which we briefly summarize.

The trace of $H^1(\Omega)$ to $\partial\Omega$ is well defined and can be identified with a Besov space $B$, the details of which will not be needed here [59]. Moreover, the kernel of the trace map is $H_0^1(\Omega)$, the $H^1$-closure of $C^\infty$ functions with compact support in $\Omega$. The domain $\mathcal{D}(\mathscr{E}_{\partial\Omega})$ can be identified with a closed subspace of $B$, and there is a bounded linear extension operator from $\mathcal{D}(\mathscr{E}_{\partial\Omega})$ to $H^1(\Omega)$, see [36]. Writing $u|_{\partial\Omega}$ for the trace of $u$ we see that the following defines a Hilbert space and inner product, see [37, Proposition 3.2].

$$W(\Omega, \partial\Omega) = \{u \in H^1(\Omega) : u|_{\partial\Omega} \in \mathcal{D}(\mathscr{E}_{\partial\Omega})\}$$
$$\langle u, v\rangle_{V(\Omega,\partial\Omega)} = \langle u, v\rangle_{H^1(\Omega)} + \mathscr{E}_{\partial\Omega}(u|_{\partial\Omega}, v|_{\partial\Omega}) + \langle u|_{\partial\Omega}, v|_{\partial\Omega}\rangle_{L^2(\partial\Omega,\mu)}.$$

One consequence of the preceding is that if we let $m = \mathcal{L}^2|_\Omega + \mu|_{\partial\Omega}$ where $\mathcal{L}^2$ is Lebesgue measure on $\Omega$ and $\mu$ is from (2.3), then for any $c_0 > 0$ the quadratic form

$$\mathscr{E}(u) := \int_\Omega (\nabla u)^2 d\mathcal{L}^2 + c_0 \mathscr{E}_{\partial\Omega}(u|_{\partial\Omega}). \tag{3.1}$$

with domain $W(\Omega, \partial\Omega))$ in $L^2(\overline{\Omega}, m)$ is a Dirichlet form.

Another consequence that will be significant in the next section is that $W(\Omega, \partial\Omega))$ may be written as the sum of $H_0^1(\Omega)$ and the subspace of $H^1(\Omega)$ obtained by the extension operator from $\mathcal{D}(\mathcal{E}_{\partial\Omega})$. In this context it is useful to describe an explicit extension operator given in [30] as the "second proof" of their Theorem 6.1, and to extract some further features of the extension and their consequences. We refer to [30] for a more detailed exposition of the construction, including diagrams of the hexagons and the triangulation.

The extension operator for a function $f$ on $\partial\Omega$ is defined as follows. There is an induction over $n = 0, 1, 2, \ldots$ that defines an exhaustion of $\Omega$ by regular hexagons of sidelength $3^{-n}$ which meet only on edges. As the induction proceeds, each new hexagon is subdivided into 6 equilateral triangles meeting at the center and a piecewise linear function is defined on each triangle by linear interpolation of the values at the vertices. The value at the center of the hexagon is defined to be the average of its vertices, and the induction is such that the hexagon vertices either lie on edges of the triangles from the previous stage of the construction, which provides the values at these points, or lie on $\partial\Omega$. In the latter case the values of $f$ are used at these vertices, and we note that those vertices of a hexagon with sidelength $3^{-n}$ that come from $\partial\Omega$ are points from the vertex set $V_{n+1}$, see Definition 2.2. The resulting piecewise linear function is the extension of $f$ to $\Omega$, which we call $g$. It is a special case of the result in [30, Theorem 6.1] that if $f$ is a piecewise harmonic function, then $g$ is Lipschitz in the Euclidean metric on $\overline{\Omega}$.

With regard to the following theorem, we note that the existence of a bounded linear extension in this setting is known [36], and we could develop the corollaries from this, but it will be useful for us to know the result for this specific extension.

**Theorem 3.1** *The extension operator described above is a bounded linear extension operator from $\mathcal{D}(\mathcal{E}_{\partial\Omega})$ with norm $(\|\cdot\|^2_{L^2(\partial\Omega,\mu)} + \mathcal{E}_{\partial\Omega}(f))^{1/2}$ to $H^1(\Omega)$ with its usual norm, and satisfies the seminorm bound*

$$\int_\Omega |\nabla g|^2 \leq C' \sum_n \sum_{x \underset{n}{\sim} y} |f(x) - f(y)|^2 \leq C\mathcal{E}_{\partial\Omega}(f). \tag{3.2}$$

We need the following elementary lemma.

**Lemma 3.2** *If $h$ is a linear function on an equilateral triangle $T \subset \mathbb{R}^2$, then $\int_T |\nabla h|^2$ over the triangle is a constant multiple of the sum of the squared differences between vertices, independent of the size of the triangle. Thus $\int |\nabla g|^2$ over a hexagon in the extension construction is bounded by a constant multiple of the squared vertex differences summed over all pairs of vertices of the hexagon.*

***Proof*** The first statement follows from a direct computation for a triangle of sidelength 1 that is left to the reader, combined with the observation that when rescaling length the scaling of $|\nabla h|^2$ cancels with that for the area of the triangle. For the second, write the integral as a sum over the triangles, use the first statement on each triangle and apply the Cauchy–Schwarz inequality. □

***Proof of Theorem 3.1*** We begin by observing that all operations in the definition of the extension are linear in $f$, and consequently so is the extension operator. In the following argument $C$ is a constant that may change value from step to step, even within an inequality.

Observe that in the construction, the vertex values of a hexagon of sidelength $3^{-n}$ come either from values at $f$ at points of $V_n$, or from linear interpolation of values of $f$ at vertices from $V_{n-1}$ on the side of a hexagon of the previous scale, or from linear interpolation between a value at a point in $V_{n-1}$ and a value obtained as an interpolant between values at vertices from $V_{n-2}$. The observation that drives the Lipschitz bound in [30] is that in any of these cases we can bound the pairwise difference between values at vertices by $C \sum |f(x) - f(y)|$ where the sum ranges over neighbor pairs in $V_n$ that are within a bounded distance from the hexagon. It follows that the number of terms in the sum has a uniform bound, and thus the squared pairwise differences are bounded by $C \sum |f(x) - f(y)|^2$. In view of Lemma 3.2, the left side of (3.2) is bounded by a constant multiple of the sum of squares over edges in the triangulation. The preceding argument gives a bound for the sum over the edges that are in a hexagon of size $3^{-n}$, and summing over the hexagons yields (3.2).

For the norm bound we must control $\|g\|_{L^2(\Omega)}$. It is immediate from the construction that $\|g\|_{L^\infty(\Omega)} \leq \|f\|_{L^\infty(\partial\Omega)}$. However the latter can be bounded in a standard manner from the resistance bound (2.2), because it implies that $|f(x)| \geq \frac{1}{2}\|f\|_\infty$ on an interval of size controlled by $\mathscr{E}_{\partial\Omega}$. Direct computation then gives $\|f\|_\infty^2 < 2(\|f\|_{L^2(\partial\Omega)}^2 + \mathscr{E}_{\partial\Omega}(f))$. Since $\|g\|_{L^2(\Omega)} \leq C\|g\|_{L^\infty(\Omega)}$ we obtain $\|g\|_{L^2(\Omega)} \leq C\|f\|_{L^2(\partial\Omega)}$. Together with the seminorm bound (3.2) this proves the extension operator is bounded. □

**Corollary 3.3** *The domain of the Dirichlet form* $\mathscr{E}(u) := \int_\Omega (\nabla u)^2 d\mathcal{L}^2 + c_0 \mathscr{E}_{\partial\Omega}(u|_{\partial\Omega})$ *may be written as a sum of compact subspaces of* $L^2(\Omega)$

$$W(\Omega, \partial\Omega) = H_0^1(\Omega) + H^1_{\mathscr{E}(\partial\Omega)}(\Omega),$$

*where* $H^1_{\mathscr{E}(\partial\Omega)}(\Omega)$ *is the image of* $\mathcal{D}(\mathscr{E}_{\partial\Omega})$ *under the extension map.*

***Proof*** The decomposition as a sum of closed subspaces follows from Wallin's result [59] that the kernel of the trace map is $H_0^1(\Omega)$, and from Lancia's bounded linear extension [36] (or from Theorem 3.1 above). Compactness of $H_0^1(\Omega)$ is from the classical Rellich–Kondrachov theorem. Compactness of $H^1_{\mathscr{E}(\partial\Omega)}(\Omega)$ is an immediate consequence of the density of the finite dimensional space of harmonic functions in $\mathcal{D}(\mathscr{E}_{\partial\Omega})$ and the boundedness of the extension map in Theorem 3.1 because it exhibits $H^1_{\mathscr{E}(\partial\Omega)}(\Omega)$ as the completion of a sequence of finite dimensional spaces in $H^1(\Omega)$. □

The following consequence is standard.

**Corollary 3.4** *The non-negative self-adjoint Laplacian L associated to the Dirichlet form by* $\mathscr{E}(u, v) = -\int_{\overline{\Omega}}(Lu)v\, dm$ *has compact resolvent and thus its spectrum is a sequence of non-negative eigenvalues accumulating only at* $\infty$.

We also note that the extension construction gives us an explicit Hölder estimate.

**Corollary 3.5** *Functions in* $H^1_{\mathscr{E}(\partial\Omega)}$ *are* $\frac{\log 2}{\log 3}$*-Hölder in the Euclidean metric on* $\overline{\Omega}$.

***Proof*** The resistance estimate (2.2) says that a function $f \in \mathcal{D}(\mathscr{E}_{\partial\Omega})$ is $\frac{1}{2}$-Hölder in the resistance metric. A pair of neighboring points in $x, y \in V_n$ are separated by resistance distance $R(x, y) \sim 4^{-n}$, so $|f(x) - f(y)| \leq C2^{-n}$ for such points. Moreover these points are separated by Euclidean distance $3^{-n}$, so $f$ is $\frac{\log 2}{\log 3}$-Hölder in the Euclidean metric.

The values of $f$ on $V_n$ are those used as boundary data on hexagons of sidelength $3^{-n}$ in the extension construction, and they obviously contribute a term with gradient bounded by $2^n 3^{-n}$. Since the extension $g$ on such a hexagon involves terms from the extensions to hexagons of scale $3^{-m}$ for $m \leq n$, we may sum these to see that $|\nabla g| \leq C2^n 3^{-n}$ on this hexagon. In particular, $|g(x) - g(y)| \leq C|x - y|^{\log 2/\log 3}$. This is true for hexagons of all scales, hence for all points in $\Omega$, and was already known for points on $\partial\Omega$, so the proof is complete. □

There are discontinuous functions in $W(\Omega, \partial\Omega)$ because it contains $H^1_0(\Omega)$. However, eigenfunctions of $\mathscr{E}$, which exist by Corollary 3.4, can be shown to be Hölder continuous.

**Theorem 3.6** *Eigenfunctions of the Dirichlet form* (3.1) *(equivalently of L as in Corollary 3.4) are Hölder continuous.*

***Sketch of the Proof*** Suppose $u$ is an eigenfunction with eigenvalue $\lambda$, so $\mathscr{E}(u, v) = -\lambda\langle u, v\rangle_{L^2(dm)}$ for all $v \in \mathcal{D}(\mathscr{E})$, so by Corollary 3.3 it is in particular true for all $v \in H^1_0(\Omega)$. We will write $\mathscr{E}_\Omega(u, v) = \int_\Omega \nabla u \cdot \nabla v\, d\mathcal{L}^2$ for the usual seminorm in $H^1(\Omega)$.

Let $h_u$ be the harmonic function on $\Omega$ with boundary data $u|_{\partial\Omega}$, where we recall that the latter denotes the trace. Since $h_u$ is the extension of $u|_{\partial\Omega}$ that minimizes $\mathscr{E}_\Omega$ and we know $u$ is an extension for which $\mathscr{E}_\Omega(u, u) < \infty$, we have $h_u \in H^1(\Omega)$ and it follows immediately that $h_u \in \mathcal{D}(\mathscr{E})$, and in fact $u - h_u \in H^1_0(\Omega)$.

Then for $v \in H^1_0(\Omega)$ we compute

$$\mathscr{E}_\Omega(u - h_u, v) = \mathscr{E}_\Omega(u, v) = \mathscr{E}(u, v) = -\lambda\langle u, v\rangle_{L^2(dm)} = -\lambda\langle u, v\rangle_{L^2(\Omega, d\mathcal{L}^2)}.$$

This uses the fact that $\mathscr{E}_\Omega(h_u, v) = 0$ because $h_u$ is harmonic and that boundary terms vanish because $u - h_u = 0$ on $\partial\Omega$ and $v|_{\partial\Omega} = 0$, the latter because $H^1_0(\Omega)$ is the kernel of the trace map.

This shows that $u - h_u \in H^1_0(\Omega)$ satisfies $\Delta_0(u - h_u) = \lambda u$, where $\Delta_0$ is the Dirichlet Laplacian. This reduces the problem to determining the regularity of the harmonic function $h_u$ and $u - h_u = \lambda G_0 u$, where $G_0$ is the Dirichlet Green's operator on $\Omega$.

It is proved in [50] that the kernel $g(x, y)$ of the Green's operator is in a Sobolev space $W^{1,q}(\Omega)$ with $q > 2$, and it follows that $g(x, y)$ is Hölder continuous. Hence so is $u - h_u = \lambda G_0 u$. The Hölder continuity depends on the regularity of the boundary $\partial\Omega$ and can, in principle, be estimated using the quantity $I_q(\Omega)$ on page 337 of [50].

The function $h_u$ is the harmonic extension of the trace $u|_{\partial\Omega}$. The latter is in $\mathcal{D}(\mathcal{E}_{\partial\Omega})$ and is therefore $\frac{\log 2}{\log 3}$-Hölder. Moreover, the Riemann mapping on $\Omega$ has Hölder continuous extension to $\partial\Omega$, see [35, 48], so $h_u$ is Hölder continuous and the proof is complete. □

*Remark 3.7* The proof of the Hölder continuity of $h_u$ discussed above admits a generalization which does not involve the Riemann mapping and is applicable in any dimension. We refer to [33, Chapter 1 Section 2] for the following reasoning; all references are to the definitions and results stated there. Our domain is class $S$ as in Definition 1.1.20, so a Lipschitz function on $\partial\Omega$ has harmonic extension in a Hölder class $C^\beta$ by Lemma 1.2.4. Taking a sequence of Lipschitz approximations to out boundary data $u|_{\partial\Omega}$ and examining the constants in the proof of Lemma 1.2.4 we discover that these blow-up in manner reflecting the Hölder continuity of $u|_{\partial\Omega}$, and in particular that $h_u$ is Hölder continuous with exponent the worse of the Hölder exponent of the boundary data and the exponent for a Lipschitz function. A particular case is Remark 1.2.5. We are indebted to Tatiana Toro for pointing out this reference and sketching the argument.

A consequence of Theorem 3.1 which will be significant later is that when high frequency oscillations on $\partial\Omega$ are extended to $\Omega$ they do not penetrate the domain very far and the energy of the extended function is very close to the energy of

**Corollary 3.8** *If the best piecewise harmonic approximations to $f \in \mathcal{D}(\mathcal{E}_{\partial\Omega})$ at scale $n$ is the zero function, then the extension $g$ is supported within distance $C_1 3^{-n}$ of the boundary and satisfies a seminorm bound of the form*

$$\int_\Omega |\nabla g|^2 \leq C_2 4^{-n} \mathcal{E}_{\partial\Omega}(f).$$

***Proof*** We saw in the construction that the values from $V_n$ were the only interpolation data for hexagons of sidelength $3^{-m}$, $m \leq n$. If $f$ is zero at these points, then $g \equiv 0$ on these hexagons, so its support is within distance at most $C_1 3^{-n}$ from $\partial\Omega$. Moreover, the terms corresponding to $V_m$, $m \leq n$ in the middle term of the seminorm bound (3.2) are zero, in which case comparing this to the definition (2.1) of the boundary energy we see that the energy scaling factor provides an additional factor of $4^{-n}$. □

## 4 Inductive Mesh Construction and Discrete Energy Forms

We inductively construct the usual approximations to the closed snowflake domain $\overline{\Omega}$, along with a triangulating mesh, in the following manner. The scale $n$ approximation consists of a collection of equilateral triangles with side length $3^{-n}$, the mesh is their edges, and those edges which belong to only one triangle are called boundary edges. When $n = 0$ there is exactly one equilateral triangle. The scale $n+1$ approximation is obtained as shown in Fig. 1: each scale $n$ triangle is subdivided into nine triangles as on the left, and triangles are appended to the centers of boundary edges as on the right. (The figure shows the case $n = 0$; at future steps at most two edges of a triangle can be boundary edges.) The $n = 2$ approximation is (f) in Fig. 2.

The mesh is treated as a graph $\Gamma_n$ with vertices $\mathcal{V}_n$. Note that the vertices $V_n$ in the graphs in Sect. 2 are a subset of $\mathcal{V}_n$ and indeed the graphs constructed there are simply the boundary subgraphs of the $\Gamma_n$. Evidently, two vertices $p, q \in \mathcal{V}_n$ are connected by an edge of the mesh if and only if $|p - q| = 3^{-n}$, where $|\cdot|$ is the Euclidean norm, in which case we write $p \sim_n q$. The vertices contained in boundary edges of $\Gamma_n$ are called boundary vertices, all other vertices are interior vertices.

Let $l(\mathcal{V}_n) = \{u \mid u : \mathcal{V}_n \to \mathbb{R}\}$ and define the graph energy of $u \in l(\mathcal{V}_n)$ by

$$\mathcal{E}_n(u) = \sum_{p_1, p_2 \in \mathcal{V}_n} c_n(p_1, p_2)(u(p_1) - u(p_2))^2, \tag{4.1}$$

where $c_n(p_1, p_2)$ is the conductance between points $p_1, p_2$, and is given by

$$c_n(p_1, p_2) = \begin{cases} 1 & \text{if } p_1 \text{ and } p_2 \text{ are connected by an interior edge,} \\ 4^n & \text{if } p_1 \text{ and } p_2 \text{ are connected by a boundary edge,} \\ 0 & \text{if } p_1 \text{ and } p_2 \text{ are not connected by an edge.} \end{cases} \tag{4.2}$$

Note that for boundary vertices both (4.2) and (4.3) agree with those in Sect. 2 and therefore the restriction of (4.1) to edges in the boundary of $\Gamma_n$ coincides with the energy defined in (2.1). Moreover, the terms corresponding to edges in $\Omega$ have weight 1, and it follows from Lemma 3.2 that for a function $g$ which is piecewise linear on equilateral triangles the sum of squared differences of the function over edges in the triangulation is a constant multiple of the Dirichlet energy $\int_\Omega |\nabla g|^2$. After computing this constant and verifying that our triangulation by $\Gamma_n$ is obtained

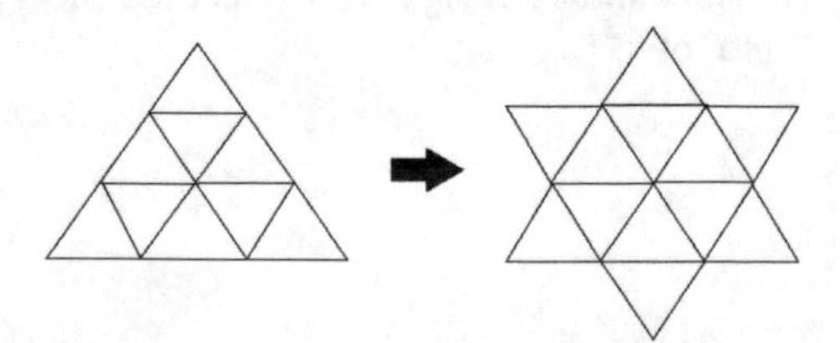

**Fig. 1** Mesh construction through scaled equilateral triangles

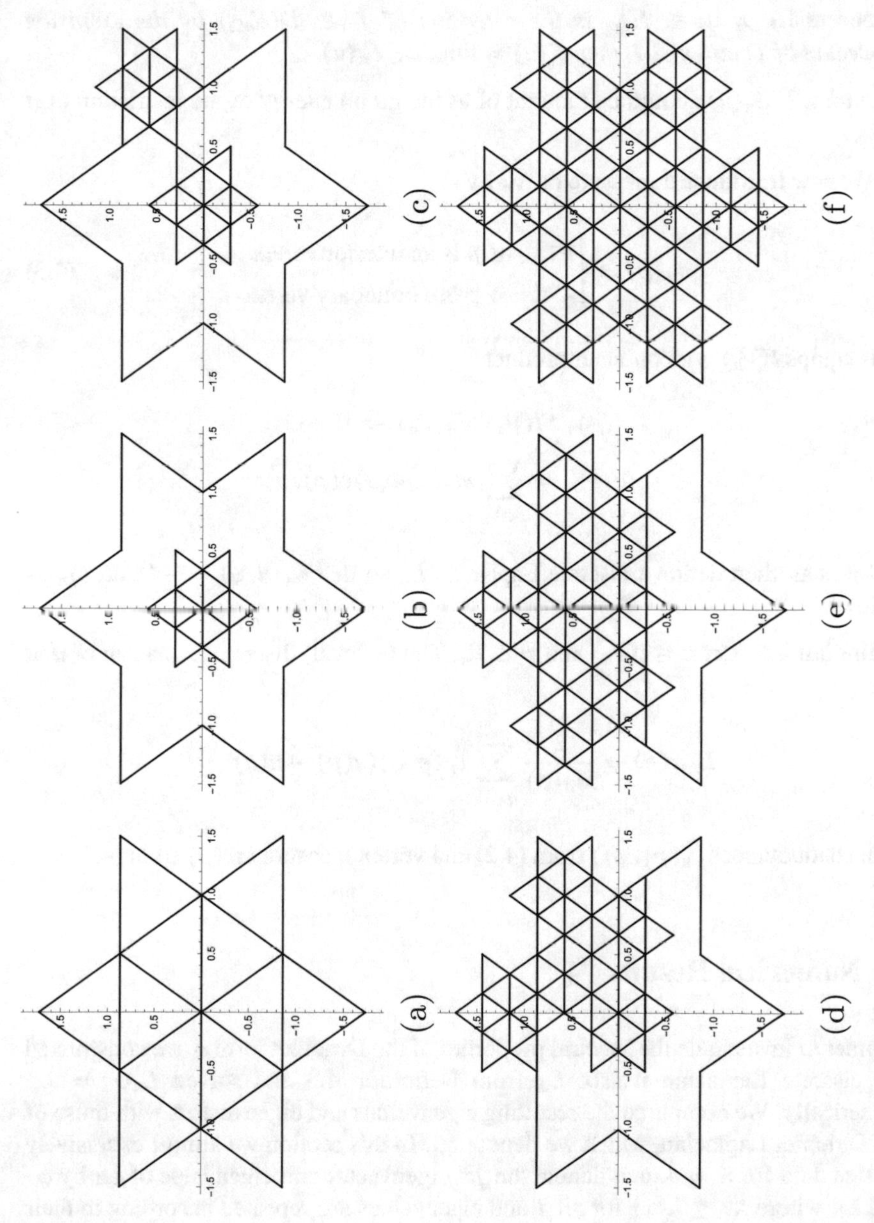


**Fig. 2** Algorithm to generate the vertices of the graph $\Gamma_n$

from the $n^{\text{th}}$ triangulation in the extension of Theorem 3.1 simply by dividing all triangles of sidelength larger than $3^{-n}$ into subtriangles of sidelength $3^{-n}$ the following consequence is immediate.

**Theorem 4.1** *If $u \in H^1_{\partial\Omega}$ is the extension of $f \in \mathcal{D}(\mathcal{E}_{\partial\Omega})$ by the extension procedure of Theorem 3.1, then $\mathcal{E}(u) = \lim_{n\to\infty} \mathcal{E}_n(u)$.*

*Remark 4.2* $\mathcal{E}_n(u)$ should be thought of as the graph energy of the restriction of $u$ to $\mathcal{V}_n$.

We now introduce a measure on $\mathcal{V}_n$ by

$$m_n(p) = \begin{cases} 9^{-n} & \text{if } p \text{ is an interior vertex} \\ 4^{-n} & \text{if } p \text{ is a boundary vertex.} \end{cases} \tag{4.3}$$

This equips $l(\mathcal{V}_n)$ with an inner product

$$\langle \cdot, \cdot \rangle_n : l(\mathcal{V}_n) \times l(\mathcal{V}_n) \to \mathbb{R}$$

$$\langle u, v \rangle_n = \sum_{p\in\mathcal{V}_n} m_n(p)u(p)v(p),$$

and we can then define a discrete Laplacian $L_n$ so that $\mathcal{E}_n(u, v) = -\langle L_n u, v\rangle_n$ as follows.

**Definition 4.3** Let $u \in l(V_n)$ and $p \in \mathcal{V}_n$. The ($n$-level) discrete Laplacian of $u$ at $p$ is

$$L_n u(p) = \frac{2}{m_n(p)} \sum_{q\in V_n} c_n(p, q)(u(p) - u(q)),$$

with conductances $c_n(p_1, p_2)$ from (4.2) and vertex measure $m_n(p)$ from (4.3).

## 5 Numerical Results

In order to investigate the spectral properties of the Dirichlet form $\mathcal{E}$ we constructed the discrete Laplacian matrix $L_n$ from Definition 4.3 and solved $L_n\phi = \lambda\phi$ numerically. We compared the resulting eigenvalues and eigenvectors with those of the Dirichlet Laplacian, which we denote $\tilde{L}_n$. In this section we almost exclusively present data for $n = 4$, and denote the $j^{\text{th}}$ eigenvector and eigenvalue of $L_4$ by $\phi_j$ and $\lambda_j$, where $\lambda_j \le \lambda_{j+1}$ for all $j$ and eigenvalues are repeated according to their multiplicity. We label the eigenvectors $\tilde{\phi}_j$ and $\tilde{\lambda}_j$ in the same manner.

In order to justify the above numerical approach to investigating the spectrum of $\mathscr{E}$ we should like to prove a result of the following kind. We believe this ought to be possible by methods analogous to those in [16, 19, 38, 39].

**Conjecture 5.1** *Our discrete and finite element approximations to eigenfunctions of $\mathscr{E}$ converge uniformly, and hence the spectrum of $L_n$ converges to that of $L$.*

## 5.1 Algorithm and Implementation

We constructed $L_n$ by an iterative procedure written in the programming language Python and then computed and graphed the eigenvalues and eigenfunctions of $L_n$ using Mathematica. We also computed the matrix $\tilde{L}_n$ of the Dirichlet Laplacian so that comparison of $L_n$ and $\tilde{L}_n$ could be used to identify effects of the boundary Dirichlet form.

The construction of $L_n$ involved constructing the mesh graphs $\Gamma_n$. The vertices of $\Gamma_1$, which is shown in Fig. 2a, were hard-coded. Construction of the vertices of $\Gamma_2$ was achieved by the steps (b)–(f) of Fig. 2: specifically, $\Gamma_1$ was scaled by $\frac{1}{3}$, translated to cover the region in the first quadrant and reflected to the other quadrants. Duplicate vertices were deleted and the process repeated to construct the vertices of $\Gamma_n$ for $n = 3$ and $n = 4$.

Construction of $L_n$ is equivalent to finding the vertex adjacencies of $\Gamma_n$; an elementary way to do this is to find pairs of points separated by distance $3^{-n}$. However it was not efficient to identify boundary vertices in this manner, and we used KD trees to improve run times for this aspect of the problem. With the adjacencies and boundary edges identified it is easy to weight these and construct the matrix $L_n$. The matrix $\tilde{L}_n$ of the Dirichlet Laplacian is obtained by deleting the rows and columns corresponding to boundary vertices.

## 5.2 The Eigenvalue Counting Function

The eigenvalue counting function for a general symmetric matrix $M$ is defined by

$$N_M(x) := \#\{\lambda \leq x \mid \lambda \text{ is an eigenvalue of } M\},$$

where eigenvalues are counted with multiplicity. This generalizes to operators with discrete spectrum in $[0, \infty)$. It is well-known that if the operator in question is the (positive) Laplacian $\Delta$ on a Euclidean domain, then $N_\Delta$ encodes a great deal of geometric information: for example, the Weyl law says that it grows as $C_d x^{d/2}$, where $d$ is the dimension and $C_d$ encodes the volume of the domain; more precise asymptotics can be obtained for certain classes of domains.

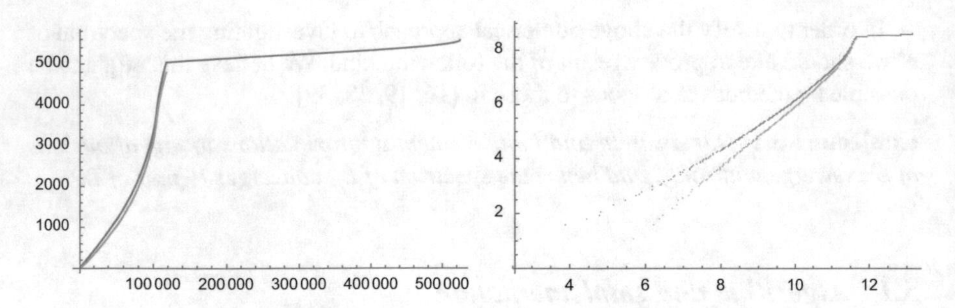


**Fig. 3** The figure on the left shows the Eigenvalue counting functions for the discrete Laplacian $L_4$ (blue) and the Dirichlet discrete Laplacian $\tilde{L}_4$ (orange) of level four. The figure on the right displays their corresponding log-log plots

**Fig. 4** The figure on the left shows the $L_4$ eigenvector $\phi_{5028}$ with the eigenvalue $\lambda_{5028} = 118038.02$. This eigenvector at the regime change $\sim 118038.5$ is qualitatively similar to the last Dirichlet eigenvector $\tilde{\phi}_{4789}$ with the eigenvalue $\tilde{\lambda}_{4789} = 118039.37$. The figure on the right shows the last Dirichlet eigenvector $\tilde{\phi}_{4789}$

The eigenvalue counting functions of $L_n$ and the Dirichlet Laplacian $\tilde{L}_n$ for the $n = 4$ approximation are displayed in Fig. 3, along with Log-Log plots that identify their growth behavior. It is immediately apparent that $N_{L4}$ has two regimes, a low eigenvalue regime in which it fairly closely tracks the behavior of $N_{\tilde{L}4}$, though with a lower initial rate of growth, and a high eigenvalue regime in which its growth is entirely different. Moreover, there is an additional feature which is not apparent from the graphs. The largest eigenvalue of $\tilde{L}_4$ is $\tilde{\lambda}_{4789} = 118{,}039.37$, and the regime-change point on the $L_4$ graph occurs at the almost identical value $\lambda_{5028} = 118{,}038.02$. Graphs of the corresponding eigenfunctions are also very similar; they are both highly oscillatory, so their graphs look multi-valued, but the similarity in the macroscopic profile of the oscillations is readily apparent in Fig. 4.

## 5.3 Eigenvectors and Eigenvalues in the Low Eigenvalue Regime

The eigenstructure of $L_n$ in the low eigenvalue regime holds few surprises. Comparing it to that for $\tilde{L}_n$ we found that the boundary Dirichlet form permitted many additional low-frequency configurations, some of which are shown in Fig. 5.

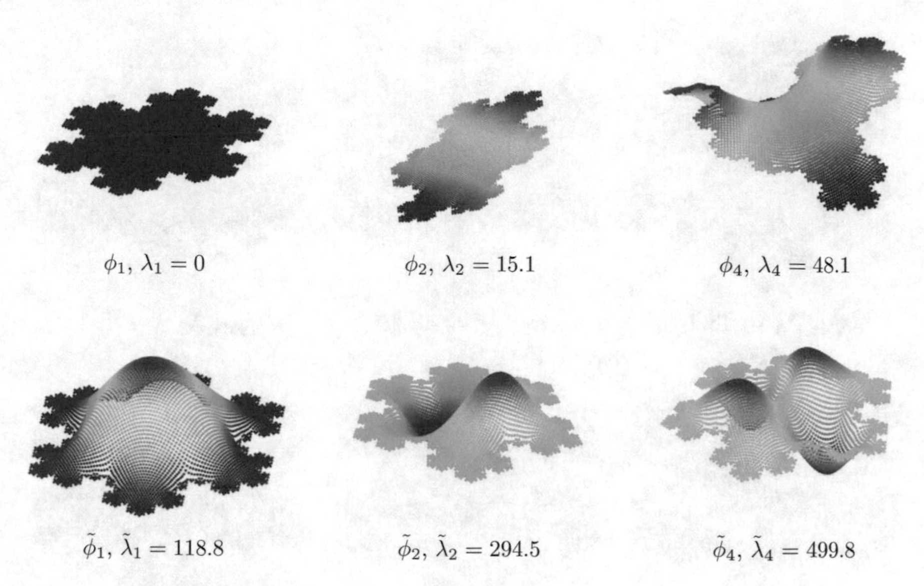


**Fig. 5** Selected eigenvectors of $L_4$ (above) and $\tilde{L}_4$ (below)

**Table 1** Eigenvalues of $L_4$ and $\tilde{L}_4$

| Eigenvalues of $L_4$ | | | | | | | | | | Eigenvalues of $\tilde{L}_4$ | | | |
|---|---|---|---|---|---|---|---|---|---|---|---|---|---|
| j | $\lambda_j$ | j | $\lambda_j$ | j | $\lambda_j$ | j | $\lambda_j$ | j | $\lambda_j$ | j | $\tilde{\lambda}_j$ | j | $\tilde{\lambda}_j$ |
| 1 | 0 | 8 | 125.4 | 15 | 344.8 | 22 | 617.9 | 29 | 880.9 | 1 | 118.8 | 8 | 822.7 |
| 2 | 15.1 | 9 | 171.6 | 16 | 363.6 | 23 | 651.6 | 30 | 880.9 | 2 | 294.5 | 9 | 822.7 |
| 3 | 15.1 | 10 | 171.6 | 17 | 482.0 | 24 | 651.6 | 31 | 1007.2 | 3 | 294.5 | 10 | 941.7 |
| 4 | 48.1 | 11 | 238.5 | 18 | 482.0 | 25 | 743.8 | 32 | 1014.9 | 4 | 499.8 | 11 | 950.5 |
| 5 | 48.1 | 12 | 238.5 | 19 | 490.9 | 26 | 787.9 | 33 | 1014.9 | 5 | 499.8 | 12 | 950.5 |
| 6 | 85.1 | 13 | 313.0 | 20 | 490.9 | 27 | 851.2 | 34 | 1098.6 | 6 | 575.1 | 13 | 1084.6 |
| 7 | 119.2 | 14 | 313.0 | 21 | 609.5 | 28 | 851.2 | | | 7 | 630.5 | | |

It is apparent from the counting function plots in Fig. 3 that this difference in the growth of $N_{L_n}$ and $N_{\tilde{L}_n}$ decreases as $\lambda$ increases.

The eigenvalues less than $\lambda = 1100$ for each operator are in the Table 1. It is evident that for both operators some eigenvalues occur with multiplicity 1 and some with multiplicity 2. This is true throughout the spectrum (not just in the low frequency regime) and is explained by symmetries of $\overline{\Omega}$; this may be verified by the argument given in [45]. At higher frequencies these symmetries are difficult to see from the graphs but readily apparent in contour plots, as illustrated in Fig. 6.

One striking observation about the spectra of $L_n$ and $\tilde{L}_n$ was the existence of eigenvectors and eigenvalues that were very similar in both. An example is shown in Fig. 7, where it is apparent that $\phi_{34}$ with eigenvalue $\lambda_{34} = 1098.6$ is very similar to $\tilde{\phi}_{13}$ with eigenvalue $\tilde{\lambda}_{13} = 1084.6$. Such "pairs" of eigenvectors, one from $L_n$ and one from $\tilde{L}_n$, both with the same symmetries and similar eigenvalues, were

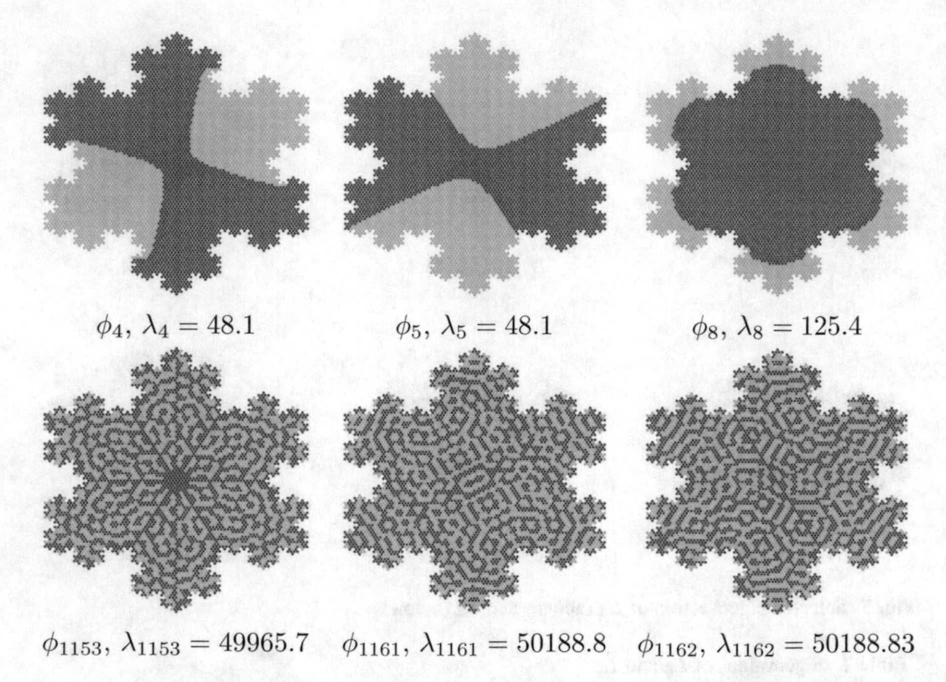


**Fig. 6** Contour plots illustrating symmetries of some eigenvectors of $L_4$. Blue indicates $|\phi| \leq \epsilon$, red is $\phi > \epsilon$, and green is $\phi < \epsilon$, for $\epsilon = 0.01$

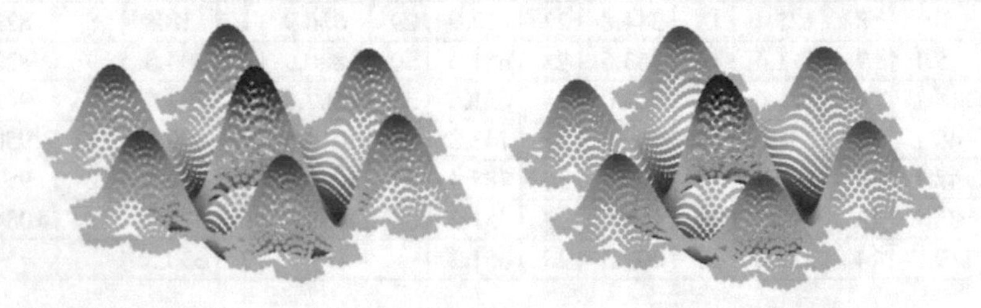

**Fig. 7** A "pair" of eigenvectors, one for $L_4$ and one for $\tilde{L}_4$, with similar symmetry structure and eigenvalues. The figure on the left shows the $L_4$ eigenvector $\phi_{34}$ with the eigenvalue $\lambda_{34} = 1098.6$. The figure on the right shows the Dirichlet eigenvector $\tilde{\phi}_{13}$ with the eigenvalue $\tilde{\lambda}_{13} = 1084.6$

found throughout the low eigenvalue regime. One possible explanation is that there might be Dirichlet eigenfunctions with symmetries that imply they are close to being eigenfunctions of $L$.

To see this, let us try to compute how far a Dirichlet eigenfunction might be from being an eigenfunction of $L$. Making a computation like that in Theorem 3.6 we see that a Dirichlet eigenfunction is $u \in H_0^1(\Omega)$ such that $\mathscr{E}_\Omega(u, v) = \lambda \langle u, v \rangle_{L^2(\Omega)}$ for all $v \in H_0^1(\Omega)$. A general element of $\mathcal{D}(\mathscr{E})$ can be written as $v + h_f$, where $f \in \mathcal{D}(\mathscr{E}_{\partial\Omega})$ denotes the boundary values and $h_f$ is harmonic. Since $u \in H_0^1(\Omega)$

has zero boundary trace we have

$$\mathscr{E}(u, v+h_f) = \mathscr{E}_\Omega(u, v) = \lambda\langle u, v+h_f\rangle_{L^2(\overline{\Omega},dm)} - \lambda\langle u, h_f\rangle_{L^2(\Omega)}$$

so $u$ would also be an eigenfunction of $L$ if $\langle u, h_f\rangle_{L^2(\Omega)} = 0$ for all $h_f$. Therefore, we would expect to find an eigenfunction and eigenvalue near $(u, \lambda)$ if $\langle u, h_f\rangle_{L^2(\Omega)}$ is small for all $f \in \mathcal{D}(\mathscr{E}_{\partial\Omega})$ normalized so that $\mathscr{E}_{\partial\Omega}(f) + \|f\|^2_{L^2(\partial\Omega)} = 1$. In consideration of Corollary 3.8 we might expect highly oscillatory $f$ to produce $h_f$ with support close to the boundary, where $u$ is small because it has Dirichlet boundary values. This heuristic suggests that we need only consider slowly varying $f$, and, moreover, that if the symmetries of $u$ are such that it is nearly orthogonal to a large enough subspace of harmonic extensions of slowly varying boundary value functions, then $u$ should be very close to being an eigenfunction of $L$ as well as $\tilde{L}$. We do not know if this heuristic is the correct explanation for the observed phenomenon, or have estimates that would make the explanation precise.

## 5.4 *Localization in the High Eigenvalue Regime*

The high eigenvalue regime contains the eigenvalues of $L_n$ that are larger than the largest Dirichlet eigenvalue. In this regime we see a dramatic transition from eigenfunctions that are supported throughout the domain to eigenfunctions localized near the boundary. Figure 8 illustrates this change using contour plots for eigenfunctions of $L_4$: the first row of images are of eigenfunctions just inside the regime, while the second row shows the progression of localization with increasing frequency. The last image is the highest eigenvalue eigenfunction of $L_4$. Note that the onset of the localization phenomenon is rapid, occurring across just a few eigenfunctions, and that continues to sharpen as the frequency increases, but at a decreasing rate.

Several phenomena in physics involve the localization of eigenmodes (e.g. Anderson localization due to a random potential in the Schrödinger equation). A nice survey from a mathematical perspective is in [49]. Some involve low frequency modes (e.g. weak localization due to irregular or complex geometry of the domain) and some involve high frequency modes (e.g. whispering-gallery modes, bouncing ball modes, and focusing modes). The type that looks most similar to those observed for $L_n$ are the whispering-gallery modes. These are well understood for convex domains [31], but despite the fact that for many years similar modes have been observed in non-convex domains, including domains with pre-fractal boundaries, our understanding of why these occur is incomplete. One approach to high frequency localized eigenmodes uses quantum billiards, but studying quantum billiards on sets with fractal boundaries seems to be a difficult problem [42]. The only literature we are aware of for the Koch snowflake domain is [41, 43], and it does not include modes of the type we see for $L_n$.

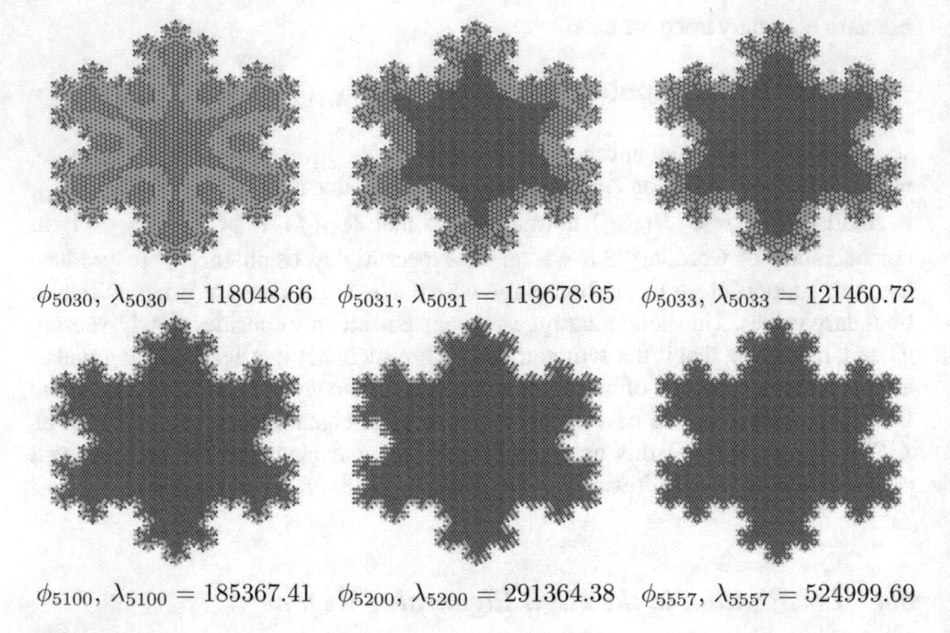


**Fig. 8** Contour plots of $L_4$ eigenvectors illustrating localization. Blue indicates $|\phi| \leq \epsilon$, red is $\phi > \epsilon$, and green is $\phi < -\epsilon$, for $\epsilon = 0.01$

We believe that in our setting the high frequency boundary localizations are not due to a whispering-gallery effect, but rather to the relatively weak coupling between the boundary energy and the domain energy. We can see this, at least heuristically, by repeating the sort of calculation done at the end of Sect. 5.3 above, but this time considering how close a high frequency eigenfunction of the boundary energy $\mathscr{E}_{\partial\Omega}$ might be to an eigenfunction of $L$. Suppose $u \in \mathcal{D}(\mathscr{E}_{\partial\Omega})$ satisfies $\mathscr{E}_{\partial\Omega}(u, v) = \lambda\langle u, v\rangle_{L^2(\Omega,\mu)}$ for all $v \in \mathcal{D}(\mathscr{E}_{\partial\Omega})$. If we extend $u$ and $v$ harmonically to $\Omega$, obtaining $h_u$ and $h_v$ respectively, and add an arbitrary $\tilde{v} \in H_0^1(\Omega)$ to $h_v$ so that $h_v + \tilde{v}$ ranges over $\mathcal{D}(\mathscr{E})$, we obtain

$$\mathscr{E}(h_u, h_v + \tilde{v}) = \mathscr{E}_{\partial\Omega}(u, v) + \mathscr{E}_{\Omega}(h_u, h_v) + \mathscr{E}_{\Omega}(h_u, \tilde{v}).$$

Now if $u$ is of high frequency it is highly oscillatory and it is not unreasonable to expect that its piecewise harmonic approximation at large scales is small. In consideration of Corollary 3.8 one might expect that then $\mathscr{E}_{\Omega}(h_u, h_u)$ is small, and therefore $\mathscr{E}(h_u, h_v + \tilde{v}) \approx \mathscr{E}_{\partial\Omega}(u, v)$. Moreover, we have

$$\langle h_u, h_v + \tilde{v}\rangle_{L^2(\overline{\Omega},dm)} = \langle h_u, h_v + \tilde{v}\rangle_{L^2(\Omega,d\mathcal{L}^2)} + \langle u, v\rangle_{L^2(\partial\Omega,d\mu)}$$

and again we see from Corollary 3.8 that one should expect $\|h_u\|_{L^2(\Omega,d\mathcal{L}^2)}$ to be small if $u$ is high frequency, leading to $\langle h_u, h_v + \tilde{v}\rangle_{L^2(\overline{\Omega},dm)} \approx \langle u, v\rangle_{L^2(\partial\Omega,d\mu)}$. Combining this with the corresponding statement for the energies and the

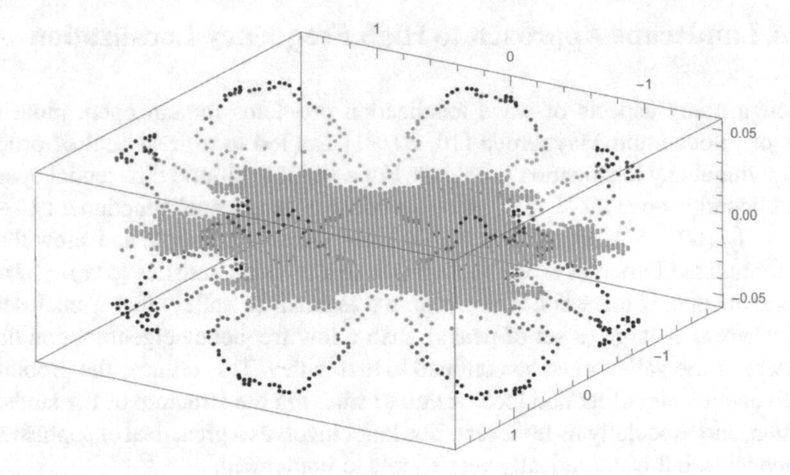


**Fig. 9** The 5550th $L_4$ eigenfunction, $\phi_{5550}$

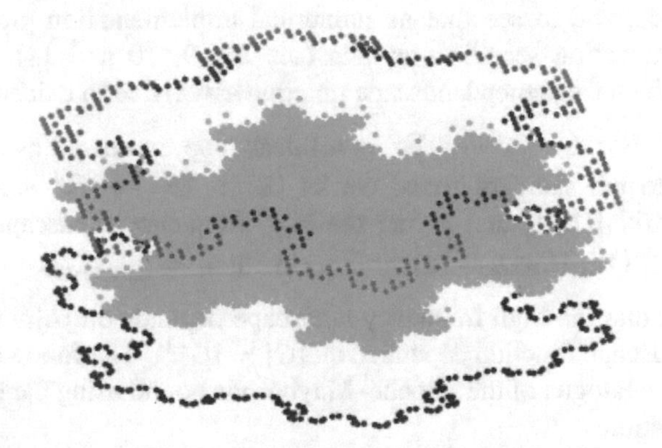

**Fig. 10** The last $L_4$ eigenfunction, $\phi_{5557}$ with $\lambda_{5557} = 524999.69$ oscillates so rapidly and is so small away from $\partial\Omega$ that its graph appears to split into copies of $\partial\Omega$ at a positive and negative height and $\Omega$ at height zero

assumption that $u$ was an eigenfunction of $\mathscr{E}_{\partial\Omega}$ we have

$$\mathscr{E}(h_u, h_v + \tilde{v}) \approx \mathscr{E}_{\partial\Omega}(u, v) = \lambda\langle u, v\rangle_{L^2(\Omega,\mu)} \approx \langle h_u, h_v + \tilde{v}\rangle_{L^2(\overline{\Omega},dm)}$$

which, if exactly true, would say that $h_u$ was an eigenfunction of $\mathscr{E}$ with eigenvalue $\lambda$. This argument is consistent with what we observe, especially the fact that oscillations of the most localized of the high frequency eigenvectors of $L_n$ appear to have wavelength approximately the length of an edge of $\Gamma_n$ and are indeed localized very closely to the boundary curve, as seen in Fig. 10. However we do not have precise arguments or estimates to justify this heuristic reasoning. A less regular looking eigenfunction localized on the boundary is shown in Fig. 9.

## 6 A Landscape Approach to High Frequency Localization

Although many aspects of wave localization problems remain open, pioneering work of Filoche and Mayboroda [10, 20, 21] has led to a great deal of progress on low-frequency localization problems. For a suitable elliptic differential operator $\mathcal{L}$ on a bounded open set $U \subset \mathbb{R}^n$, they introduce a "landscape" function $u : U \to \mathbb{R}$ by $u = \int_U |G(x, y)|$, where $G$ is the Dirichlet Green's function, and show that an $L^\infty$ normalized Dirichlet eigenfunction $\phi$ with $\mathcal{L}\phi = \lambda\phi$ satisfies $|\phi(x)| \leq \lambda u(x)$. In consequence, if there is a region where $u$ is small (a valley) that is enclosed by a set where $u$ is large (a set of peaks), then a low frequency eigenfunction that is non-zero in the valley must be confined to that valley. This reduces the problem of low frequency eigenfunction localization to studying the structure of the landscape function, and especially its level sets. The latter involves a great deal of sophisticated mathematics, but is numerically very simple to implement.

Our purpose here is to write a high frequency variant of the argument of Filoche and Mayboroda, and to see that its numerical implementation predicts the high frequency localization seen in our data (see Fig. 9, 10 and 11). We do so by introducing a high frequency landscape function associated to a linear map $L$.

**Definition 6.1** For $L : \mathbb{R}^d \to \mathbb{R}^d$ is a linear map expressed as a matrix $\left[L_{ij}\right]$ with respect to the standard basis, we let $|L| : \mathbb{R}^d \to \mathbb{R}^d$ denote the linear map with matrix $\left[|L_{ij}|\right]$ and define the high frequency landscape vector to be $u = |L|(1, \cdots, 1)^T$. Thus $u = (u_1, \cdots, u_d)$ with $u_i = \sum_j |L_{ij}|$.

We remark that the high frequency landscape depends on $|L|$ whereas the low frequency landscape function depended on $|G| = |\mathcal{L}^{-1}|$. The following elementary theorem is the analogue of the Filoche–Mayboroda bound using the high frequency landscape function.

**Theorem 6.2** *Let $\phi = (\phi_1, \ldots, \phi_d)^T \in \mathbb{R}^d$ be an eigenvector of $L$ corresponding to an eigenvalue $\lambda > 0$ normalized such that $\max_{1 \leq i \leq d} |\phi_i| = 1$. Then for each $i = 1, \cdots, d$,*

$$\phi_i \leq \frac{1}{\lambda} u_i. \tag{6.1}$$

***Proof*** Since $\lambda\phi = L\phi$ we may compute using the standard basis $\{e_i\}$

$$\lambda\phi_i = \langle e_i, L\phi\rangle = \sum_j \phi_j \langle e_i, L\phi_j\rangle = \sum_j \phi_j L_{ij} \leq \sum_j |L_i j| = u_i,$$

where the inequality used the normalization that $|\phi_j| \leq 1$ for all $j$. □

The result of numerical computation of the high frequency landscape function for $L = L_4$, our discrete Laplacian, is shown in Fig. 11. It appears to be constant

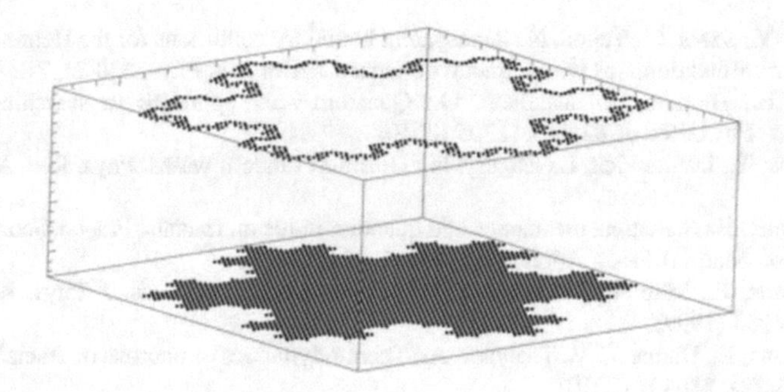

**Fig. 11** The high frequency landscape for $L_4$

on $\Omega$ and on $\partial\Omega$, but in fact we find that it attains two values on $\partial\Omega$. This can be computed directly from the formula in Definition 4.3.

**Lemma 6.3** *The high frequency landscape function u for $L_n$ has constant value $8 \cdot 3^{2n+1}$ on $\Omega$ and attains the two values $2^{4n+3}$ and $2^{4n+3} + 3 \cdot 2^{2n+2}$ on $\partial\Omega$.*

In consideration of the $L^\infty$ normalization of the eigenvector $\phi$ the following is immediate from the lemma and Theorem 6.2.

**Corollary 6.4** *The constraint* (6.1) *is effective in $\Omega$ if $\lambda \geq 8 \cdot 3^{2n+1}$.*

Observe that for $L_4$ this constraint is $\lambda \geq 157464$. As expected, this is somewhat higher than the value $\lambda = 118038.02$ at which we identified the regime change in the spectrum, because we did not have complete localization at the regime change.

*Remark 6.5* It should be noted that Theorem 6.2 works well in our setting simply because the values in the matrix $L_n$ are very different on boundary edges than on interior edges. This is the same reason that drove our heuristic reasoning in Sect. 5 regarding the decoupling of the interior and boundary energies and their effect on the spectrum of $L$. The high frequency landscape approach would not be expected to predict whispering-gallery modes, or localized modes due to quantum scarring, as these are not due solely to local properties of the Laplacian.

**Acknowledgments** The authors are grateful to Michael Hinz, Maria Rosaria Lancia, Svitlana Mayboroda, Anna Rozanova-Pierrat, and Tatiana Toro for helpful discussions.

The authors are grateful to Kevin Marinelli for the support in implementing the numerical codes.

Research supported in part by NSF DMS Grants 1659643 and 1613025.

## References

1. Achdou, Y., Sabot, C., Tchou, N.: Diffusion and propagation problems in some ramified domains with a fractal boundary. M2AN Math. Model. Numer. Anal. **40**(4), 623–652 (2006)

2. Achdou, Y., Sabot, C., Tchou, N.: Transparent boundary conditions for the Helmholtz equation in some ramified domains with a fractal boundary. J. Comput. Phys. **220**(2), 712–739 (2007)
3. Agliari, E., Blumen, A., Muelken, O.: Quantum-walk approach to searching on fractal structures. Phys. Rev. A **82**(1), 012305 (2010)
4. Aharonov, Y., Davidovich, L., Zagury, N.: Quantum random walks. Phys. Rev. A **48**(2), 1687 (1993)
5. Akkermans, E.: Statistical mechanics and quantum fields on fractals. Fractal Geom. Dyn. Syst. Pure Appl. Math. II Fract. Appl. Math. **601**, 1–22 (2013)
6. Akkermans, E., Mallick, K.: Vortices in Ginzburg-Landau billiards. J. Phys. A Math. Gen. **32**(41), 7133 (1999)
7. Akkermans, E., Dunne, G.V., Teplyaev, A.: Thermodynamics of photons on fractals. Phys. Rev. Lett. **105**(23), 230407 (2010)
8. Aougab, T., Dong, S.C., Strichartz, R.S.: Laplacians on a family of quadratic Julia sets II. Commun. Pure Appl. Anal. **12**(1), 1–58 (2013)
9. Apushkinskaya, D.E., Nazarov, A.I.: A survey of results on nonlinear Venttsel problems. Appl. Math. **45**(1), 69–80 (2000)
10. Arnold, D.N., David, G., Filoche, M., Jerison, D., Mayboroda, S.: Localization of eigenfunctions via an effective potential. Commun. Partial Differ. Equ. **44**(11), 1186–1216 (2019)
11. Attal, S., Petruccione, F., Sabot, C., Sinayskiy, I.: Open quantum random walks. J. Stat. Phys. **147**(4), 832–852 (2012)
12. Baelus, B.J., Cabral, L.R.E., Peeters, F.M.: Vortex shells in mesoscopic superconducting disks. Phys. Rev. B **69**(6), 064506 (2004)
13. Brzoska, A., Coffey, A., Hansalik, M., Loew, S., Rogers, L.G.: Spectra of magnetic operators on the diamond lattice fractal (2017, preprint). arXiv:1704.01609
14. Capitanelli, R., Vivaldi, M.A.: Uniform weighted estimates on pre-fractal domains. Discrete Contin. Dynam. Syst.-Ser. B **19**(7), 1969–1985 (2014)
15. Cefalo, M., Lancia, M.R.: An optimal mesh generation algorithm for domains with Koch type boundaries. Math. Comput. Simul. **106**, 133–162 (2014)
16. Creo, S., Lancia, M.R., Vélez-Santiago, A., Vernole, P.: Approximation of a nonlinear fractal energy functional on varying Hilbert spaces. Commun. Pure Appl. Anal. **17**(2), 647–669 (2018)
17. Creo, S., Lancia, M.R., Vernole, P., Hinz, M., Teplyaev, A.: Magnetostatic problems in fractal domains (2018, preprint ). arXiv:1805.08262
18. Dunne, G.V.: Heat kernels and zeta functions on fractals. J. Phys. A: Math. Theor. **45**(37), 374016 (2012)
19. Evans, E.J.: A finite element approach to Hölder extension using prefractals. Methods Appl. Anal. **19**(2), 161–186 (2012)
20. Filoche, M., Mayboroda, S.: The hidden landscape of localization (2011). arXiv:1107.0397
21. Filoche, M., Mayboroda, S.: Universal mechanism for Anderson and weak localization. Proc. Natl. Acad. Sci. USA **109**(37), 14761–14766 (2012)
22. Fleckinger, J., Levitin, M., Vassiliev, D.: Heat equation on the triadic von Koch snowflake: asymptotic and numerical analysis. Proc. Lond. Math. Soc. **71**(2), 372–396 (1995)
23. Flock, T.C., Strichartz, R.S.: Laplacians on a family of quadratic Julia sets I. Trans. Am. Math. Soc. **364**(8), 3915–3965 (2012)
24. Freiberg, U.R., Lancia, M.R.: Energy form on a closed fractal curve. Z. Anal. Anwendungen **23**(1), 115–137 (2004)
25. Geim, A.K., Dubonos, S.V., Grigorieva, I.V., Novoselov, K.S., Peeters, F.M., Schweigert, V.A.: Non-quantized penetration of magnetic field in the vortex state of superconductors. Nature **407**(6800), 55 (2000)
26. Grebenkov, D.S., Nguyen, B.-T.: Geometrical structure of Laplacian eigenfunctions. SIAM Rev. **55**(4), 601–667 (2013)
27. Hinz, M., Rogers, L.: Magnetic fields on resistance spaces. J. Fractal Geom. **3**(1), 75–93 (2016)
28. Hinz, M., Teplyaev, A.: Dirac and magnetic Schrödinger operators on fractals. J. Funct. Anal. **265**(11), 2830–2854 (2013)

29. Hinz, M., Koch, D., Meinert, M.: Sobolev spaces and calculus of variations on fractals (2018, preprint). arXiv:1805.04456
30. Hinz, M., Lancia, M.R., Teplyaev, A., Vernole, P.: Fractal snowflake domain diffusion with boundary and interior drifts. J. Math. Anal. Appl. **457**(1), 672–693 (2018)
31. Keller, J.B., Rubinow, S.I.: Asymptotic solution of eigenvalue problems. Ann. Phys. **9**(1), 24–75 (1960)
32. Kempe, J.: Quantum random walks: an introductory overview. Contemp. Phys. **44**(4), 307–327 (2003)
33. Kenig, C.E.: Harmonic Analysis Techniques for Second Order Elliptic Boundary Value Problems. CBMS Regional Conference Series in Mathematics, vol. 83. Published for the Conference Board of the Mathematical Sciences, Washington; by the American Mathematical Society, Providence (1994)
34. Kigami, J.: Analysis on Fractals. Cambridge Tracts in Mathematics, vol. 143. Cambridge University Press, Cambridge (2001)
35. Koskela, P., Reitich, F.: Hölder continuity of Sobolev functions and quasiconformal mappings. Math. Z. **213**(3), 457–472 (1993)
36. Lancia, M.R.: A transmission problem with a fractal interface. Z. Anal. Anwendungen **21**(1), 113–133 (2002)
37. Lancia, M.R., Vernole, P.: Venttsel' problems in fractal domains. J. Evol. Equ. **14**(3), 681–712 (2014)
38. Lancia, M.R., Vélez-Santiago, A., Vernole, P.: Quasi-linear Venttsel' problems with nonlocal boundary conditions on fractal domains. Nonlinear Anal. Real World Appl. **35**, 265–291 (2017)
39. Lancia, M.R., Vélez-Santiago, A., Vernole, P.: A quasi-linear nonlocal Venttsel' problem of Ambrosetti-Prodi type on fractal domains. Discrete Contin. Dyn. Syst. **39**(8), 4487–4518 (2019)
40. Lapidus, M.L.: Fractal drum, inverse spectral problems for elliptic operators and a partial resolution of the Weyl-Berry conjecture. Trans. Am. Math. Soc. **325**(2), 465–529 (1991)
41. Lapidus, M.L., Niemeyer, R.G.: Towards the Koch snowflake fractal billiard: computer experiments and mathematical conjectures. In: Gems in Experimental Mathematics. Contemporary Mathematics, vol. 517, pp. 231–263. American Mathematical Society, Providence (2010)
42. Lapidus, M.L., Niemeyer, R.G..: The current state of fractal billiards. In: Fractal Geometry and Dynamical Systems in Pure and Applied Mathematics. II. Fractals in Applied Mathematics. Contemporary Mathematics, vol. 601, pp. 251–288. American Mathematical Society, Providence (2013)
43. Lapidus, M.L., Niemeyer, R.G.: Sequences of compatible periodic hybrid orbits of prefractal Koch snowflake billiards. Discrete Contin. Dyn. Syst. **33**(8), 3719–3740 (2013)
44. Lapidus, M.L., Pang, M.M.H.: Eigenfunctions of the Koch snowflake domain. Commun. Math. Phys. **172**(2), 359–376 (1995)
45. Lapidus, M.L., Neuberger, J.W., Renka, R.J., Griffith, C.A.: Snowflake harmonics and computer graphics: numerical computation of spectra on fractal drums. Int. J. Bifur. Chaos Appl. Sci. Eng. **6**(7), 1185–1210 (1996)
46. Mosco, U.: Composite media and asymptotic Dirichlet forms. J. Funct. Anal. **123**(2), 368–421 (1994)
47. Mosco, U.: Variational fractals. Annali Della Scuola Normale Superiore Di Pisa. Classe di Scienze **25**(3–4), 683–712 (1997)
48. Näkki, R., Palka, B.: Extremal length and Hölder continuity of conformal mappings. Comment. Math. Helv. **61**(3), 389–414 (1986)
49. Nguyen, B.-T., Grebenkov, D.S.: Localization of Laplacian eigenfunctions in circular, spherical, and elliptical domains. SIAM J. Appl. Math. **73**(2), 780–803 (2013)
50. Nyström, K.: Integrability of Green potentials in fractal domains. Ark. Mat. **34**(2), 335–381 (1996)
51. Ord, G.N.: Fractal space-time: a geometric analogue of relativistic quantum mechanics. J. Phys. A Math. Gen. **16**(9), 1869 (1983)

52. Rogers, L.G., Teplyaev, A.: Laplacians on the basilica Julia sets. Commun. Pure Appl. Anal. **9**(1), 211–231 (2010)
53. Sapoval, B.: Experimental observation of local modes in fractal drums. Phys. D Nonlinear Phenomena **38**(1), 296–298 (1989)
54. Sapoval, B., Gobron, T.: Vibrations of strongly irregular or fractal resonators. Phys. Rev. E **47**, 3013–3024 (1993)
55. Sapoval, B., Gobron, T., Margolina, A.: Vibrations of fractal drums. Phys. Rev. Lett. **67**(21), 2974 (1991)
56. Spicer, C., Strichartz, R.S., Totari, E.: Laplacians on Julia sets III: Cubic Julia sets and formal matings. In: Fractal Geometry and Dynamical Systems in Pure and Applied Mathematics. I. Fractals in Pure Mathematics. Contemporary Mathematics, vol. 600, pp. 327–348. American Mathematical Society, Providence (2013)
57. Strichartz, R.S., Wiese, S.C.: Spectral properties of Laplacians on snowflake domains and filled Julia sets (2019). arXiv:1903.08259
58. Ventcel', A.D.: On boundary conditions for multi-dimensional diffusion processes. Theor. Probab. Appl. **4**, 164–177 (1959)
59. Wallin, H.: The trace to the boundary of Sobolev spaces on a snowflake. Manuscripta Math. **73**(2), 117–125 (1991)

# On the Dimension of the Sierpinski Gasket in $l^2$

M. V. Marchi

**Abstract** We prove that, in the Hilbert space $l^2$, there exists exactly one closed and bounded set *invariant* with respect to a countable family $\mathscr{S} = \{ \mathrm{S}_k : k \geq 0\}$ of contractive similarities whose fixed points are the vertices of an infinite dimensional unitary simplex.

Such a set $K$, called the *infinite dimensional Sierpinski gasket*, turns out to have *infinite generalized Hausdorff dimension*.

We also prove that, for any given countable set of weights $\mathcal{W} = \{m_k : k \geq 0\}$, there exists exactly one probability measure $\mu$ invariant with respect to $\mathscr{S}$ and $\mathcal{W}$, i.e. such that $\mu = \sum_{k=0}^{+\infty} m_k \, \mu \circ \mathrm{S}_k^{-1}$ in the weak convergence.

$\mu$, which is supported in $K$, results also to be the limit of the sequence of probability measures invariant with respect to the families of the first $N+1$ similarities $\mathscr{S}_N$ and of weights $\mathcal{W}_N = \{\frac{m_k}{\sum_{i=0}^{N} m_i} : 0 \leq k \leq N\}$.

**Keywords** 28A80 · 28A33 · 28C05 · 46B20 · 46E27

## 1 Introduction

In the Hilbert space $l^2$, we consider the countable family $\mathscr{S} = \{ \mathrm{S}_k : k \geq 0\}$ of the contractive similarities

$$\mathrm{S}_k(p) = p(k) + \frac{1}{2}(p - p(k)),$$

whose fixed points $p(k)$, defined in the Sect. 2.1, are the vertices of an infinite dimensional unitary simplex.

M. V. Marchi (✉)
Dip. di Matematica, Sapienza Università di Roma, Roma, Italy
e-mail: marchi@mat.uniroma1.it

 
M. R. Lancia, A. Rozanova-Pierrat (eds.), *Fractals in Engineering: Theoretical Aspects and Numerical Approximations*, SEMA SIMAI Springer Series 8,
https://doi.org/10.1007/978-3-030-61803-2_5

It is well-known [4] that, for any $N \in \mathbb{N}$, there exists exactly one closed and bounded set *invariant* with respect to the family of the first N+1 similarities $\mathscr{S}_N = \{\mathrm{S}_0, \mathrm{S}_1, \ldots, \mathrm{S}_N\}$, i.e. a set $K_N$, called the *N-dimensional Sierpinski gasket*, such that

$$K_N = \bigcup_{k=0}^{N} \mathrm{S}_k(K_N).$$

$K_N$ is a compact subset of $\mathbb{R}^N$, whose Hausdorff dimension $D_H(K_N) = \frac{\log(N+1)}{\log 2}$ is strictly less than $N$.

In order to find a closed set invariant with respect to $\mathscr{S}$, i.e. such that

$$K = \bigcup_{k \geq 0} \mathrm{S}_k(K), \tag{1}$$

we cannot apply, as in the case of finitely many similarities, the Banach fixed point theorem to the map $\boldsymbol{\mathcal{S}}(A) = \bigcup_{k \geq 0} \mathrm{S}_k(A)$, defined on $\mathcal{H} = \{A \subset X : A \neq \emptyset$, closed and bounded$\}$, because there exist closed sets whose image fails to be closed.

Nevertheless, the map $\overline{\boldsymbol{\mathcal{S}}} : \mathcal{H} \to \mathcal{H}$ defined by $\overline{\boldsymbol{\mathcal{S}}}(A) = \overline{\bigcup_{k \in \mathbb{N}} \mathrm{S}_k(A)}$, is a contraction on $\mathcal{H}$, that, once equipped with the Hausdorff distance, is a complete metric space.

Then, there exists exactly one closed and bounded set $K$ such that

$$K = \overline{\bigcup_{k \geq 0} \mathrm{S}_k(K)}. \tag{2}$$

and at most one *invariant* with respect to $\mathscr{S}$. Actually, any closed set that satisfies (1) must also satisfy (2).

We prove that such an invariant set exists, by proving that $\overline{\bigcup_{N \in \mathbb{N}} K_N}$ is invariant with respect to $\boldsymbol{\mathcal{S}}$.

We are interested to know if $K$, which fails to be compact, has some intermediate properties between the compactness and the separability of $l^2$.

An in-between condition is to have, for any $\delta > 0$, a *fine-δ-cover*. This condition is equivalent to the possibility to provide $K$ with a *finite generalized Hausdorff dimension*, i.e. to find some *Hausdorff measure function h* such that $\Lambda_h(K) = 0$, where $\Lambda_h(K)$ is the $h$-Hausdorff measure of $K$ [3].

Because of its completeness and uniform spread in infinite directions, the set $K$ does not have *fine-δ-covers* for all $\delta > 0$. Then, it has *infinite generalized Hausdorff dimension*, that, differently from $K_N$ with respect to $\mathbb{R}^N$, is the "same Hausdorff dimension" of the space $\boldsymbol{l}^2$ [2], in which it is included.

Finally, in analogy to the finite case [4], we prove that, for any given countable set of weights

$$\mathcal{W} = \{m_k : k \geq 0,\ m_k > 0,\ \sum_{k=0}^{+\infty} m_k = 1\},$$

there exists exactly one probability measure $\mu$ *invariant* with respect to $\mathscr{S}$ and $\mathcal{W}$, i.e. such that

$$\mu = \sum_{k=0}^{+\infty} m_k \mu \circ \mathrm{S}_k^{-1} \tag{3}$$

in the weak convergence.

In order to prove the existence of such a measure, we introduce in $\mathcal{M}^1$, the spaces of probability measures with bounded support, the Kantorovich metric

$$\delta(\mu, \nu) = \sup\{|\int f d\mu - \int f d\nu| :\ f \in \mathrm{Lip}(X),\ L_f \leq 1\}, \tag{4}$$

that metrizes the weak topology.

We prove that $(\mathcal{M}^1, \delta)$ is a complete metric space and that the sequence $(\mathcal{T}_N)_{N \in \mathbb{N}}$ of contractions on $\mathcal{M}^1$, defined by

$$\mathcal{T}_N(\mu) = \sum_{k=0}^{N} m_k^N \mu \circ \mathrm{S}_k^{-1},$$

where

$$m_0^N = \frac{m_0}{\sum_{i=0}^{N-1} m_i}, \ldots, m_{N-1}^N = \frac{m_{N-1}}{\sum_{i=0}^{N-1} m_i},$$

converges uniformly to a contraction $\mathcal{T}$.

The fixed point of $\mathcal{T}$ satisfies (3) and is supported in the Sierpinski gasket $K$. It is also the limit of the sequence of the fixed points of $\mathcal{T}_N$.

In [5] it is proved the existence of an invariant separable measure that satisfies (3).

In Sect. 2, we prove that the infinite dimensional Sierpinski gasket exists and has *infinite generalized dimension*. In Sect. 3, we prove the existence and the properties of the measure $\mu$ invariant with respect to $\mathscr{S}$ and $\mathcal{W}$.

## 2 Invariant Sets

Let $X$ be the Hilbert space $l^2$. We denote by $\cdot$ the inner product, by $|\ \ |$ the norm, and by $d$ the Hilbert metric. For $r > 0$, $p \in X$ and $E \subset X$, we set $B(p, r) = \{q \in X : d(p, q) < r\}$ and $B(E, r) = \{q \in X : d(q, E) < r\}$.

Let $p(0) = 0$, $p(1) = e_1$, and, for $k \geq 2$, $p(k) = (x(k)_1, x(k)_2, \ldots)$, where

$$x(k)_i = \begin{cases} \frac{1}{\sqrt{2}} \frac{1}{\sqrt{i}\sqrt{i+1}} & i < k \\ \frac{1}{\sqrt{2}} \frac{\sqrt{k+1}}{\sqrt{k}} & i = k \\ 0 & i > k. \end{cases}$$

$p(0), p(1), \ldots$. are the vertices of an *infinite dimensional unitary simplex*. Actually, for any $j, k$ and $j \neq k$,

$$d(p(j), p(k)) = 1 \tag{5}$$

and then, for any $k > 0$, $|p(k)| = 1$. Moreover, for any $i, j, k$ mutually distinct,

$$(p(i) - p(j)) \cdot (p(k) - p(j)) = \frac{1}{2}.$$

We will denote by $V = \{p(0),\ p(1),\ \ldots\}$.

For any $p(k)$, let $\mathrm{S}_k : X \to X$ be the contractive similarity of contraction factor $\frac{1}{2}$ and fixed point $p(k)$

$$\mathrm{S}_k(p) = p(k) + \frac{1}{2}(p - p(k)) = \frac{1}{2}(p + p(k)).$$

In the following, for any $j, k$ we set $p(j, k) = \mathrm{S}_j(p(k)) = \mathrm{S}_k(p(j))$. From (5) it follows that for any $i, j, k$ mutually distinct

$$d(p(i, j), p(i, k)) = d(\mathrm{S}_i(p(j)), \mathrm{S}_i(p(k)) = \frac{1}{2}. \tag{6}$$

For any $\{i_1, i_2 \ldots, i_n\} \in \{0, 1, 2, \ldots\}^n$, we set
$\mathrm{S}_{i_1 i_2 \ldots i_n} = \mathrm{S}_{i_1} \circ \mathrm{S}_{i_2} \circ \ldots \circ \mathrm{S}_{i_n}$ and, for any $B \subset X$, $B_{i_1 i_2 \ldots i_n} = \mathrm{S}_{i_1 i_2 \ldots i_n}(B)$.

## 2.1 Infinite Dimensional Sierpinski Gasket

Let $\mathcal{H} = \{A \subset X : A \neq \emptyset$, closed and bounded$\}$ be equipped with the Hausdorff distance

$$\rho(A, B) = \sup\{d(p, B), d(q, A) : p \in A, q \in B\},$$

where $d(p, E) = \inf\{d(p, q) : q \in E\}$. $(\mathcal{H}, \rho)$ is a complete metric space.

For any $A \in \mathcal{H}$, let us define

$$\boldsymbol{\mathcal{S}}(A) = \bigcup_{k\geq 0} \mathrm{S}_k(A) \quad \text{and} \quad \overline{\boldsymbol{\mathcal{S}}}(A) = \overline{\bigcup_{k\geq 0} \mathrm{S}_k(A)}.$$

**Definition 2.1** $A$ is said to be invariant with respect to the family $\mathscr{S}$, if

$$A = \boldsymbol{\mathcal{S}}(A).$$

In order to prove the existence of closed and bounded invariant sets, we cannot apply the Banach fixed point theorem to the map $\boldsymbol{\mathcal{S}}$, since $\boldsymbol{\mathcal{S}}[\mathcal{H}] \nsubseteq \mathcal{H}$. Indeed, for instance, the image $\boldsymbol{\mathcal{S}}(D)$ of the bounded and closed set $D = \{(-p(k) + \frac{1}{k}e_k : k \geq 2\}$ contains the sequence $(\mathrm{S}_k(-p(k) + \frac{1}{k}e_k))_{k\geq 2} = (\frac{1}{2k}e_k)_{k\geq 2}$, which converges to $p(0) \notin \boldsymbol{\mathcal{S}}(D)$.

Instead, $\overline{\boldsymbol{\mathcal{S}}}$ is a contraction on $(\boldsymbol{H}, \rho)$. Actually, $\overline{\boldsymbol{\mathcal{S}}}[\mathcal{H}] \subseteq \mathcal{H}$ and

$$\rho(\overline{\boldsymbol{\mathcal{S}}}(A), \overline{\boldsymbol{\mathcal{S}}}(B)) = \rho(\overline{\bigcup_{k\geq 0} \mathrm{S}_k(A)}, \overline{\bigcup_{k\geq 0} \mathrm{S}_k(B)}) = \rho(\bigcup_{k\geq 0} \mathrm{S}_k(A), \bigcup_{k\geq 0} \mathrm{S}_k(B)) \leq$$

$$\sup{}_{k\geq 0}\, \rho(\mathrm{S}_k(A), \mathrm{S}_k(B)) = \frac{1}{2}\rho(A, B).$$

Therefore, by the Banach fixed point theorem, there exists exactly one $K \in \boldsymbol{H}$ such that

$$K = \overline{\boldsymbol{\mathcal{S}}}(K). \tag{7}$$

We will prove that $K$ is also invariant with respect to $\mathscr{S}$ and that it is the only element of $\mathcal{H}$ with such a property.

As in finite dimensional case, there are bounded sets invariant with respect to $\mathscr{S}$. For example

**Proposition 2.1** $K_\infty = \bigcup_{N\in\mathbb{N}} K_N$ *is invariant with respect to* $\mathscr{S}$.

***Proof*** Let $p \in \bigcup_{j\in\mathbb{N}} \mathrm{S}_j(K_\infty)$ and let $k$, $n$ in $\mathbb{N}$ be such that $p \in \mathrm{S}_k(K_n)$. We can assume $k \leq n$ otherwise, $K_n \subset K_k$, and then we choose $n = k$. From $K_n = \bigcup_{j\leq n} \mathrm{S}_j(K_n)$ it follows that $p \in K_n \subset K_\infty$.

On the other side, let $p \in K_\infty$, and let $n$ be such that $p \in K_n$. From $K_n = \bigcup_{j \le n} S_j(K_n)$, it follows that $p \in S_k(K_n)$, for some $k \le n$. Then, $p \in \bigcup_{j \le n} S_j(K_n) \subseteq \bigcup_{j \in \mathbb{N}} S_j(K_\infty)$. □

$K_\infty$ is bounded, since it is included in $\overline{B}(0, 1)$, but, as we will prove in the next subsection, it fails to be closed.

We prove that $\overline{K}_\infty$ too is invariant with respect to $\mathscr{S}$.

**Lemma 2.1** *For any bounded set A invariant with respect to* $\mathscr{S}$,

*(i)* $\boldsymbol{S}(\overline{A}) \subseteq \overline{A}$;
*(ii)* $\overline{A}$ *satisfies (7);*
*(ii)* $\overline{A} = K$.

***Proof***

(i)

$$\bigcup_{k \ge 0} S_k(\overline{A}) = \bigcup_{k \ge 0} \overline{S_k(A)} \subseteq \overline{\bigcup_{k \ge 0} S_k(A)} = \overline{A}.$$

(ii)

$$\overline{A} = \overline{\bigcup_{k \ge 0} S_k(A)} \subseteq \overline{\bigcup_{k \ge 0} S_k(\overline{A})}.$$

On the other side, from i) it follows $\overline{\bigcup_{k \ge 0} S_k(\overline{A})} \subseteq \overline{A}$.

(iii) $\overline{A} = K$, because of the uniqueness of the fixed point of $\boldsymbol{S}$.

□

By (i) of the previous lemma and from Proposition 2.1, to prove that $\overline{K}_\infty$ is invariant, it sufficies to prove the inclusion $\overline{K}_\infty \subseteq \boldsymbol{S}(\overline{K}_\infty)$.

In proving the last inclusion, we use the following Lemma 2.2. This lemma is a corollary of Theorem 2.3, which is proved in the last subsection, since it involves some properties of the N-dimensional simplices.

**Lemma 2.2** *There exists* $c > 0$, *such that, for any* $N \in \mathbb{N}$, $j, k \le N$, $j \ne k$, $p \in S_j(K_N)$, *and* $p' \in S_k(K_N)$,

$$d(p, p(j, k)) \le c\, d(p, p').$$

**Proposition 2.2** $\overline{K}_\infty \subseteq \boldsymbol{S}(\overline{K}_\infty)$.

***Proof*** Let $p \in \overline{K}_\infty$ and $(p_n)_{n \in \mathbb{N}} \subset K_\infty = \bigcup_{k \ge 0} S_k(K_\infty)$ be a sequence that converges to $p$.

We claim that there exist at most two different copies $S_j(K_\infty)$, $S_k(K_\infty)$ that contain infinitely many elements of the sequence $(p_n)_{n \in \mathbb{N}}$.

Actually, for a fixed $0 < \varepsilon < \frac{1}{4c}$, let $\overline{n}$ be sufficiently large so that $d(p_{\overline{n}}, p_m) < \varepsilon$ for any $m \geq \overline{n}$.

Let $k_{\overline{n}}, k_m, j_{\overline{n}}, j_m, N$ be such that $p_{\overline{n}} \in \mathrm{S}_{k_{\overline{n}}}(K_{j_{\overline{n}}})$, $p_m \in \mathrm{S}_{k_m}(K_{j_m})$, and $N = \max\{k_{\overline{n}}, k_m, j_{\overline{n}}, j_m\}$, then $p_{\overline{n}} \in \mathrm{S}_{k_{\overline{n}}}(K_N)$, and $p_m \in \mathrm{S}_{k_m}(K_N)$.

By Lemma 2.2, if $k_m \neq k_{\overline{n}}$,

$$d(p_{\overline{n}}, p(k_{\overline{n}}, k_m)) \leq cd(p_{\overline{n}}, p_m) < c\varepsilon.$$

Therefore, if there existed $m_1 > \overline{n}$ and $m_2 > \overline{n}$, such that $k_{m_1}, k_{m_2}, k_{\overline{n}}$ were mutually distinct, then, from the last inequality it would follow

$$d(p(k_{\overline{n}}, k_{m_1}), p(k_{\overline{n}}, k_{m_2}))) \leq d(p_{\overline{n}}, p(k_{\overline{n}}, k_{m_1})) + d(p_{\overline{n}}, p(k_{\overline{n}}, k_{m_2})) < 2c\varepsilon < \frac{1}{2},$$

which contradicts (6).

In the case of $p_n \in \mathrm{S}_j(K_\infty)$ for infinitely many $n$ and $p_n \in \mathrm{S}_k(K_\infty)$ for infinitely many $n$, $p = p(j, k) \in K_\infty = \mathcal{S}(K_\infty) \subset \mathcal{S}(\overline{K}_\infty)$.

Actually, for any $\varepsilon > 0$, let $\overline{n}$ be sufficiently large so that $d(p_n, p_m) < \varepsilon$ for any $n, m \geq \overline{n}$. For any $n \geq \overline{n}$, let $m \geq \overline{n}$ be such that $p_m \in \mathrm{S}_k(K_\infty)$ if $p_n \in \mathrm{S}_j(K_\infty)$ and $p_m \in \mathrm{S}_j(K_\infty)$ if $p_n \in \mathrm{S}_k(K_\infty)$, then, by Lemma 2.2,

$$d(p_n, p(j, k))) \leq cd(p_n, p_m) < c\varepsilon.$$

In the case of $p_n \in \mathrm{S}_j(K_\infty)$ definitively, let $p_n = \mathrm{S}_j(q_n)$. The sequence $(q_n)_{n\in\mathbb{N}} = (\mathrm{S}_j^{-1}(p_n))_{n\in\mathbb{N}} = (2p_n - p(j))_{n\in\mathbb{N}}$ is a Cauchy sequence in $K_\infty$, then, by the completeness of $X$, it converges to $q \in \overline{K}_\infty$. Therefore $p_n = \mathrm{S}_j(q_n) \to \mathrm{S}_j(q)$. By uniqueness of the limit, $p = \mathrm{S}_j(q) \in \mathcal{S}(\overline{K}_\infty)$. □

From the invariance of $\overline{K}_\infty$ and (iii) of Lemma 2.1 it follows that

**Theorem 2.1** *$K$ is invariant with respect to $\mathscr{S}$. Moreover it is the only non-empty closed bounded set invariant with respect to $\mathscr{S}$.*

## 2.2 Hausdorff Dimension of Invariant Sets

A *Hausdorff function* is a continuous increasing function $h : [0, a[\to \mathbb{R}$, positive for $t > 0$ and such that $h(0) = 0$.

Given a Hausdorff function $h$, the $h$-Hausdorff measure on $X$ is the Borel measure $\Lambda^h$ defined, for any Borel subset $B$ of $X$, by

$$\Lambda^h(B) = \lim_{\delta \to 0+} \Lambda^h_\delta(B) \quad \text{where} \quad \Lambda^h_\delta(B) = \inf\{\sum_{i=1}^{+\infty} h(\operatorname{diam} U_i) : B \subseteq \bigcup_{i=1}^{+\infty} U_i\}$$

and the infimum is taken over all countable $\delta$-covers of $B$.

We recall that a $\delta$-cover is a cover whose elements have diameter less than $\delta$.

**Definition 2.2** $B \subseteq X$ is said to have finite generalized Hausdorff dimension, if $\Lambda^h(B) = 0$ for some Hausdorff measure function $h$, otherwise it is said to have infinite generalized Hausdorff dimension.

**Definition 2.3** A countable cover $(U_i)_{i\in\mathbb{N}}$ of $B \subseteq X$ is said to be a fine-cover if $\lim_{i\to+\infty} \operatorname{diam} U_i = 0$.

A fine-repeated cover is a fine-cover such that $B \subseteq \bigcup_{i\geq k} U_i$ for every $k \in \mathbb{N}$.

It is easy to see that $B$ has a fine-$\delta$-cover for any $\delta > 0$, iff it has a fine-repeated cover.

**Proposition 2.3 ([3])** *$B \subseteq X$ has finite generalized Hausdorff dimension iff it has a fine-repeated cover.*

***Proof*** Let $B$ have finite generalized Hausdorff dimension, then $\Lambda^h(X) = 0$, for some Hausdorff measure function $h$. Hence, for any $\delta > 0$, it is possible to choose a $\delta$-cover $(U_i)_{i\in\mathbb{N}}$ such that $\sum_{i=1}^{+\infty} h(\operatorname{diam} U_i)$ converges.

This convergence, along with monotonicity of $h$, implies $\lim_{i\to+\infty} \operatorname{diam} U_i = 0$.

Conversely, if $(U_i)_{i\in\mathbb{N}}$ is a fine-repeated cover of $B$, it is possible to choose a Hausdorff measure function $h$ such that $\sum_{i=1}^{+\infty} h(\operatorname{diam} U_i)$ converges.

For any $\delta > 0$, let $\overline{n}(\delta)$ be such $\operatorname{diam} U_i < \delta$ for all $i \geq \overline{n}(\delta)$. Then $\Lambda^h_\delta(B) < \sum_{i=\overline{n}(\delta)}^{+\infty} h(\operatorname{diam} U_i)$. From the convergence of $\sum_{i=1}^{+\infty} h(\operatorname{diam} U_i)$ it follows that $\lim_{\delta\to 0^+} \Lambda^h_\delta(B) = 0$. □

To calculate the dimension of $K$, we remark that $\operatorname{diam} K = 1$, since $\operatorname{diam} K_N = 1$ for any $N \in \mathbb{N}$, and then $\operatorname{diam} K_\infty = 1$. We also remark that $K$ cannot have *finite $\delta$-covers*, whenever $\delta < 1$, since $V \subset K$. Otherwise, it would be $d(p(j), p(k)) < \delta$ for some (in fact infinitely many) $j \neq k$.

Therefore, for any $(i_1 i_2, .., i_n)$, $\operatorname{diam} K_{i_1 i_2 \dots i_n} = 2^{-n}$ and $K_{i_1 i_2 .. i_n}$ cannot have finite $\delta$-cover, whenever $\delta < 2^{-n}$, because it contains $V_{i_1 i_2 .. i_n}$, whose infinite elements have with one another distance $2^{-n}$.

Moreover, from the invariance of $K$, it also follows that, for any $n, m \in \mathbb{N}$,

$$K = \mathcal{S}^n(K) = \bigcup_{(i_1 i_2 \dots i_n)\in\{0,1,2\dots\}^n} K_{i_1 i_2 \dots i_n},$$

$$K_{i_1 i_2 \dots i_n} = \bigcup_{(i_{n+1} i_{n+2} \dots i_{n+m})\in\{0,1,2\dots\}^m} K_{i_1 i_2 \dots i_n i_{n+1} i_{n+2} \dots i_{n+m}}.$$

**Proposition 2.4** *For any $n \in \mathbb{N}$, $(i_1 \dots i_n) \in \{0, 1, 2 \dots\}^n$ and every finite family of balls $B(q_1, r_1), \dots, B(q_N, r_N)$ with $r_i < 2^{-n-3}$, there exist $i_{n+1}, i_{n+2}, i_{n+3}$, such that*

$$K_{i_1 i_2 .. i_n i_{n+1} i_{n+2} i_{n+3}} \bigcap \Big(\bigcup_{i=1}^{N} B(q_i, r_i)\Big) = \emptyset.$$

***Proof*** If there existed $N$ such that for any $i_{n+1}, i_{n+2}, i_{n+3} \in \{0, 1, 2 \ldots\}^3$ it were $K_{i_1 i_2 .. i_n i_{n+1} i_{n+2} i_{n+3}} \bigcap (\bigcup_{i=1}^{N} B(q_i, r_i)) \neq \emptyset$, then $(B(q_i, 2^{-n-2}))_{i=1,2,..N}$ would be a finite $2^{-n-1}$—cover of $K_{i_1 i_2 \ldots i_n}$ and this contradicts the previous remark. □

**Theorem 2.2** *$K$ has infinite generalized Hausdorff dimension.*

***Proof*** By Proposition 2.3, in order to prove that $K$ has the infinite generalized dimension, it is sufficient to prove that it fails to have fine-repeated cover.

Assume there exists a fine-repeated cover $(U_i)_{i \in \mathbb{N}}$. Without loss of generality, we can assume $U_i$ to be open balls $B(q_i, r_i)$, with $r_i < 2^{-3}$.

Let $n_1$ be such that $r_i < 2^{-6}$ for every $i > n_1$. By Proposition 2.4, there exist $i_1, i_2, i_3$ such that

$$K_{i_1 i_2 i_3} \bigcap (\bigcup_{i=1}^{n_1} B(q_i, r_i)) = \emptyset.$$

Let $n_2$ be such that $r_i < 2^{-9}$ for every $i > n_2$. By Proposition 2.4, there exist $i_4, i_5, i_6$ be such that

$$K_{i_1 i_2 \ldots i_6} \bigcap (\bigcup_{i=n_1}^{n_2} B(q_i, r_i)) = \emptyset.$$

By proceeding in the same way, we find a decreasing sequence of closed bounded sets

$$K_{i_1 \ldots i_{3n}} \subset K_{i_1 \ldots i_{3(n-1)}}.$$

From the completeness of $X$, it follows that there exists $p \in \cap K_{i_1 .., i_{3n}}$. By construction, $p \notin \bigcup_{i=1}^{+\infty} B(q_i, r_i)$. □

We conclude by observing that $K_\infty \neq \overline{K}_\infty$. Actually $\overline{K}_\infty = K$ has infinite generalized Hausdorff dimension, whereas $K_\infty$, that fails to have finite dimension, being $\sigma$-compact has fine-repeated covers and then finite generalized Hausdorff dimension.

## 2.3 N-Dimensional Simplices

In this section, $\mathbb{R}^N$ denotes the subspace of $l^2$ generated by $e_1, e_2, \ldots, e_N$, where $e_i$ are the elements of the Hilbert basis.

For any $N \in \mathbb{N}$,

$$T_N = \{\sum_{i=0}^{N} \alpha_i p(i) : \sum_{i=0}^{N} \alpha_i = 1, \ \alpha_i \geq 0\}$$

is the N-dimensional simplex of vertices $p(0), p(1), \dots p(N)$.

$T_N$ is a closed and convex subset of $\mathbb{R}^N$.

Since for any $N \in \mathbb{N}$, $K_N \subset T_N$, Lemma 2.2 follows from the next Theorem 2.3.

**Theorem 2.3** *There exists $c > 0$, such that, for any $N \in \mathbb{N}$, $j, k \leq N$, $j \neq k$, $p \in S_j(T_N)$, and $p' \in S_k(T_N)$,*

$$d(p, p(j,k)) \leq c\, d(p, p'). \tag{8}$$

We will prove (8) by using the law of sines, then we will need to control from below, uniformly with respect to $N$, the angles between two line segments that belong to two different copies $S_j(T_N)$ and $S_k(T_N)$. Such a lower bound is given in Proposition 2.5. In the Lemmas 2.3 and 2.4 we summarize some geometrical properties of $T_N$ that we will use in proving the Proposition.

In all this section, $j, k \leq N$ and $j \neq k$. For any $j$ and $k$,

$$T^j = \{p \in T_N : \alpha_j = 0\}$$

is the (N-1)-dimensional simplex called *facet of $T_N$ opposite to $p(j)$*, and

$$T^{j,k} = T^{kj} = T^j \cap T^k = \{p \in T_N : \alpha_j = \alpha_k = 0\}$$

is the (N-2)-dimensional simplex intersection of two facets.

We denote by $l_N$ the length of the apothems, i.e. the distance of the vertices $p(j)$ of a facet $P^k$ from the opposite N-2-dimensional simplices $T^{j,k}$, and by $q^k(j)$ the point of $T^{jk}$ such that $d(p(j), q^k(j)) = l_N$.

The unitary vector

$$u^k(j) = \frac{q^k(j) - p(j)}{d(q^k(j), p(j))}$$

has the direction of the apothem from $p(j)$ of the facet $T^k$ and is orthogonal to $T^{j,k}$.

We denote by $\widehat{vw} = \arccos \frac{v \cdot w}{|v||w|}$ the convex non-oriented angle between two vectors $v$ and $w$. For any three points $p, q, r \in \mathbb{R}^N$, we set $\widehat{pqr}$ for $\widehat{(p-q)(r-q)}$.

In $\mathbb{R}^N$ we consider the standard orientation, i.e. we say that an ordered basis $\{v_1, v_1, \dots, v_N\}$ is in the positive orientation if the determinant of the matrix $\left(v_1 |\, v_2 | \dots v_N |\right)$ is positive. The ordered set of the vertices $\{p(1), p(2), \dots, p(N)\}$ of $T_N$ respects this orientation.

Given an oriented (N-1)-dimensional subspace $U \subset \mathbb{R}^N$ with an ordered basis $\{u_1, u_2, \dots, u_{N-1}\}$ in the positive orientation of $U$, the unit normal vector $n_U$ to $U$ is chosen so that $(u_1, u_2, \dots, u_{N-1}, n_U)$ is in the positive orientation of $\mathbb{R}^N$.

**Definition 2.4** The dihedral angle between two oriented (N-1)-dimensional subspaces $U$ and $W$ of $\mathbb{R}^N$ is the angle between $n_U$ and $n_W$, chosen as above.

The dihedral angle between oriented hyperplanes is the dihedral angle between the subspaces which are translates of the hyperplanes and the dihedral angle between two facets of $T_N$ is the dihedral angle between the hyperplanes containing the facets, oriented consistently with the orientation of the facets, i.e., the order of their vertices.

We will denote by $n_j$ the unit vector orthogonal to the facet $T^j$ and by $\alpha_N$ the measure of the dihedral angle between two facets.

By the symmetry of $T_N$, neither $l_N$ nor $\alpha_N$ depends on vertices and facets. Then, without loss of generality, we sketch the proof of the Lemma 2.3 by using the facets $T^N$ and $T^{N-1}$.

The reason of this choice is that the hyperplane that contains $T^N$ is the space $\mathbb{R}^{N-1}$ and the N-2-dimesional plane that contains $T^{N-1,N}$ is the space $\mathbb{R}^{N-2}$, so that

$$l_N = d(p(N-1), T^{N-1,N}) = d(p(N-1), q^N(N-1)) = \frac{\sqrt{N}}{\sqrt{2}\sqrt{N-1}},$$

where $q^N(N-1) = (x_1, x_2, \dots x_N)$ and $x_i = \begin{cases} \frac{1}{\sqrt{2}\sqrt{i}\sqrt{i+1}} & i \le N-2 \\ 0 & N-1 \le i \le N \end{cases}$.

Moreover, since $T^k = \mathscr{R}^k_j(T^j)$, where $\mathscr{R}^k_j$ is the affine rotation in $\mathbb{R}^N$ which leaves unchanged the elements of the N-2-dimensional plane containing $T^{jk}$ and maps $p(k)$ in $p(j)$, we choose $j = N$ and $k = N-1$, because $\mathscr{R}^{N-1}_N$ turns out to be linear and the matrix $\mathscr{A}^{N-1}_N$, which represents $\mathscr{R}^{N-1}_N$ with respect to the standard basis, has the simple form

$$\mathscr{A}^{N-1}_N = \begin{pmatrix} 1 \; 0 \; 0 & \dots & 0 & 0 \\ 0 \; 1 \; 0 & \dots & 0 & 0 \\ .. \; .. \; .. & \dots & .. & 0 \\ 0 \; \dots \dots & 1 & 0 & 0 \\ 0 \; \dots \dots & 0 & \frac{1}{N} & -\frac{\sqrt{N^2-1}}{N} \\ 0 \; \dots \dots & 0 & \frac{\sqrt{N^2-1}}{N} & \frac{1}{N} \end{pmatrix}.$$

$\frac{1}{N}$ and $\frac{\sqrt{N^2-1}}{N}$ are respectively the cosine and the sine of the angle of this rotation, which is, in fact $\alpha_N$. Indeed, $n_N = e_N$ and $n_{N-1} = \mathscr{R}^{N-1}_N(e_N) = (0, \dots, 0, -\frac{\sqrt{N^2-1}}{N}, \frac{1}{N})$ as well as $u^N(N-1) = -e_{N-1}$ and $u^{N-1}(N) = \mathscr{R}^{N-1}_N(-e_{N-1}) = (0, \dots, -\frac{1}{N}, -\frac{\sqrt{N^2-1}}{N})$. Then we proved

**Lemma 2.3**

*(i)* $l_N = \frac{\sqrt{N}}{\sqrt{2}\sqrt{N-1}}$;

*(ii)* $\cos\alpha_N = \frac{1}{N}$;

*(iii) for any $j$, $k$*

$$\cos \alpha_N = u^k(j) \cdot u^j(k). \tag{9}$$

**Lemma 2.4** *For any $j$, $k$*

*(i) for any $q \in T^{jk}$,*

$$\sin(\widehat{vu^k(j)}) \leq \frac{\sqrt{N-2}}{\sqrt{2}\sqrt{N-1}}, \tag{10}$$

*where $v = (q - p(j))$;*

*(ii) for any $q$, $q' \in T^{jk}$*

$$\sin(\widehat{vw}) \geq \sqrt{1 - (\frac{N^2-2}{2N(N-1)})^2}, \tag{11}$$

*where $v = (q - p(j))$ and $w = (q' - p(k))$.*

***Proof***

(i)

$$\cos(\widehat{vu^k(j)}) = \frac{l_N}{d(p(j), q)} \geq l_N,$$

since $d(p(j), q) \leq d(p(j), p(i)) = 1$ for any $i \neq j$. Then,

$$\sin(\widehat{vu^k(j)}) \leq \sqrt{1 - l_N^2} = \frac{\sqrt{N-2}}{\sqrt{2}\sqrt{N-1}}.$$

(ii) We decompose $\frac{v}{|v|}$ and $\frac{w}{|w|}$ in their projections parallel and orthogonal respectively to $u^k(j)$ and $u^j(k)$:

$$\frac{v}{|v|} = \cos(\widehat{vu^k(j)})\, u^k(j) + \sin(\widehat{vu^k(j)})\, u_1$$

$$\frac{w}{|w|} = \cos(\widehat{wu^j(k)})\, u^j(k) + \sin(\widehat{wu^j(k)})\, u_2,$$

where $u_1$ and $u_2$ are unitary vectors in $T^{j,k}$. Then, by a rough estimate, from (9) and (10), it follows

$$|\cos(\widehat{vw})| = \frac{|u \cdot v|}{|u||w|} =$$

$$|(\cos(\widehat{vu^k(j)})\, u^k(j) + \sin(\widehat{vu^k(j)})\, u_1) \cdot (\cos(\widehat{wu^j(k)})\, u^j(k) + \sin(\widehat{wu^j(k)})\, u_2)| =$$

$$|\cos(\widehat{vu^k(j)}) \cos(\widehat{wu^j(k)})\, (u^k(j) \cdot u^j(k)) + \sin(\widehat{vu^k(j)}) \sin(\widehat{wu^j(k)})\, (u_1 \cdot u_2)| \leq$$

$$u^k(j) \cdot u^j(k) + \sin(\widehat{vu^k(j)}) \sin(\widehat{wu^j(k)}) \leq \frac{1}{N} + \frac{1}{2}\frac{N-2}{N-1} = \frac{N^2-2}{2N(N-1)}.$$

Therefore

$$\sin(\widehat{vw}) \geq \sqrt{1 - (\frac{N^2-2}{2N(N-1)})^2}.$$

□

**Proposition 2.5** *There exists $\tilde{c} > 0$, such that, for any $N \in \mathbb{N}$, any $j, k \leq N$, $j \neq k$, $p \in S_j(T^j k)$ and $p' \in S_k(T^j k)$, $p \neq p(j,k)$ and $p' \neq p(j,k)$,*

$$\sin(\widehat{pp(jk)p'}) \geq \tilde{c}. \tag{12}$$

***Proof*** Let $q, q' \in T^{jk}$ be such that $p = S_j(q')$ and $p' = S_k(q)$. The vectors $(p - p(j,k)) = (S_j(q') - S_j(p(k))) = S_j(v)$ and $(p' - p(j,k)) = (S_k(q) - S_k(p(j))) = S_k(w)$ are parallel, respectively, to the vectors $w = (q' - p(k))$ and $v = (q - p(j))$. Then, the inequality (11) is satisfied also by the angle $\widehat{pp(jk)p'}$.

(12) follows from the estimate of the minimum value of the sequence $\left(\sqrt{1 - (\frac{N^2-2}{2N(N-1)})^2}\right)_{N \geq 2}$, which is $\tilde{c} = \frac{\sqrt{95}}{12}$. □

***Proof of Theorem 2.3*** The copies $S_j(T_N)$ and $S_k(T_N)$ belong to the opposite half-spaces with respect to the hyperplane containing the facet $S_j(T^j)$, then, for any two points $p \in S_j(T_N)$, and $p' \in S_k(T_N)$, the line segment between $p$ and $p'$ must intersect $S_j(T^j)$ in a point that we denote by $q$. By the same argument, the line segment between $q$ and $p'$ must intersect $S_k(T^k)$ in a point that we denote by $q'$.

When $q = p(j,k)$, also $q' = p(j,k)$ and then $d(p, p(j,k)) \leq d(p, p')$.

When $q \neq p(j,k)$, also $q' \neq p(j,k)$. In this case,

$$d(p,q) + d(q,q') + d(q',p') = d(p,p'). \tag{13}$$

From the law of sines and (12), it follows

$$d(q,(p(j,k))) = \frac{\sin(\widehat{qq'p(j,k))}}{\sin \widehat{qp(j,k)q'})} d(q,q') \leq c\,d(q,q'), \tag{14}$$

where $c = \tilde{c}^{-1} > 1$. Because of the triangular inequality $d(p, p(j,k)) \leq d(p,q) + d(q, p(j,k))$, from (14) and (13), it follows

$$d(p, p(j,k)) \leq d(p,q) + c\,d(q,q') \leq c\,d(p,p').$$

□

## 3 Invariant Measures

Let $\mathcal{M}^e$ be the space of all finite Borel measures on X with bounded support and $\mathcal{M}^1 = \{\mu \in \mathcal{M}^e : \mu(X) = 1\}$. We denote by $\mathcal{B}(X)$ the $\sigma$-algebra of the Borel sets.

Let $\mathcal{C}^b(X)$ be the family of bounded continuous functions on $X$. For $f \in \mathcal{C}^b(X)$, we set $||f|| = \sup_{p \in X} |f(p)|$.

Given a *Stone vector lattice* $\mathcal{V} \subseteq \mathcal{C}^b(X)$, i.e. a vector lattice that satisfies the Stone condition

$$\inf(f,1) \in \mathcal{V} \qquad \text{for all } f \in \mathcal{V},$$

we say that $A \subseteq X$ is $\mathcal{V}$*-open* if there exists an increasing sequence $(u_n)_{n \in \mathbb{N}}$ of positive functions in $\mathcal{V}$ such that $1_A = \sup u_n$.

In the following, we will consider the Stone vector lattice $\mathcal{L} = \{f \in \mathcal{C}^b(X) : \mathrm{Lip} f < +\infty\}$ and its subset $\mathcal{L}_1 = \{f \in \mathcal{L} : \mathrm{Lip} f \leq 1\}$.

Given a countable set of weights $\mathcal{W} = \{m_i : i \geq 0,\ m_i > 0,\ \sum_{i=0}^{+\infty} m_i = 1\}$, for every $N \geq 0$, consider the finite set of weights $\mathcal{W}_N = \{m_0^N, m_1^N, \ldots, m_N^N\}$, where, for any $i \leq N$,

$$m_i^N = \frac{m_i}{\sum_{k=0}^N m_k}.$$

Let us remark that for any $N \in \mathbb{N}$, $j > 0$, $i \leq N$

$$m_i^N - m_i = (1 - \sum_{k=0}^{N} m_k) m_i^N \qquad m_i^N - m_i^{N+j} = m_i^N \sum_{k=N+1}^{N+j} m_k^{N+j}. \tag{15}$$

Define $\mathcal{Z}_N : \mathcal{M}^1 \to \mathcal{M}^e$ and $\mathcal{T}_N : \mathcal{M}^1 \to \mathcal{M}^1$, by

$$\mathcal{Z}_N(\mu) = \sum_{i=0}^{N} m_i \mu \circ \mathrm{S}_i^{-1} \qquad \mathcal{T}_N(\mu) = \sum_{i=0}^{N} m_i^N \mu \circ \mathrm{S}_i^{-1}.$$

For any $\mu \in \mathcal{M}^1$, we denote by $\sum_{i=0}^{+\infty} m_i \mu \circ \mathrm{S}_i^{-1}$ the weak limit of the sequence $(\mathcal{Z}_N(\mu))_{N\in\mathbb{N}}$, and we say that $\mu$ is invariant with respect to $\mathscr{S}$ and $\mathcal{W}$ if

$$\mu = \sum_{i=0}^{+\infty} m_i \mu \circ \mathrm{S}_i^{-1}.$$

From (15) it follows

**Proposition 3.1** *$(\mathcal{Z}_N(\mu))_{N\in\mathbb{N}}$ converges weakly to a measure in $\mathcal{M}^1$ iff $(\mathcal{T}_N(\mu))_{N\in\mathbb{N}}$ converges weakly to the same measure.*

***Proof*** Actually, for any $f \in \mathcal{C}^b(X)$,

$$|\int f\, d\mathcal{T}_N(\mu) - \int f\, d\mathcal{Z}_N(\mu)| =$$

$$(1-\sum_{k=0}^{N} m_k)|\sum_{i=0}^{N} m_i^N \int f \circ \mathrm{S}_i\, d\mu| \le (1-\sum_{k=0}^{N} m_k)\,||f||.$$

Letting $N \to +\infty$, we obtain the required result. □

In order to find a measure invariant with respect to $\mathscr{S}$ and $\mathcal{W}$, we will consider the sequence $(\mathcal{T}_N(\mu))_{N\in\mathbb{N}}$ in $\mathcal{M}^1$, instead of $(\mathcal{Z}_N(\mu))_{N\in\mathbb{N}}$ in $\mathcal{M}^e$.

Actually, once introduced in the space $\mathcal{M}^1$ the *Kantorovich metric* $\delta$:

$$\delta(\mu, \nu) = \sup\{|\int f\, d\mu - \int f\, d\nu| : f \in \mathcal{L}_1\}, \qquad \text{for } \mu, \nu \in \mathcal{M}^1$$

which takes into account the metric $d$ of $X$, involved in the similarities $\mathrm{S}_i$, we can prove that, for all $\mu \in \mathcal{M}^1$ the sequence $(\mathcal{T}_N(\mu))_{N\in\mathbb{N}}$ converges, in the metric $\delta$, to a measure $\mathcal{T}(\mu) \in \mathcal{M}^1$ and that $\mathcal{T} : \mathcal{M}^1 \to \mathcal{M}^1$, the pointwise limit of $(\mathcal{T}_N(\mu))_{N\in\mathbb{N}}$, is a contraction on the complete metric space $(\mathcal{M}^1, \delta)$.

Therefore, there exists a fixed point of $\mathcal{T}$, i.e. a measure that we denote by $\mu_\infty$ in $\mathcal{M}^1$, such that

$$\lim_{N\to+\infty} \delta(\mathcal{T}_N(\mu_\infty), \mu_\infty) = 0.$$

Since the convergence in the metric $\delta$ of a sequence $(\mu_n)_{n\in\mathbb{N}}$ to a measure $\mu$ implies that the real sequence $(\int f\, d\mu_n)_{n\in\mathbb{N}}$ converges to $\int f\, d\mu$ for all $f \in \mathcal{L}$, the weak convergence of $(\mathcal{T}_N(\mu))_{N\in\mathbb{N}}$ to $\mathcal{T}t(\mu)$ follows from *portmanteau* theorem that characterizes the weak convergence by weak convergence with respect to Lipschitz functions only:

**Lemma 3.1** *(see, f.i. [7, Thm.7.5.] or [6, Thm.6.1.]) For any sequence $(\mu_n)_{n\in\mathbb{N}}$ in $\mathcal{M}^1$, the following statements are equivalent:*

(*i*) $\lim_{n\to+\infty} \int f\, d\mu_n = \int f\, d\mu$ *for all* $f \in \mathcal{C}^b(X)$
(*ii*) $\lim_{n\to+\infty} \int f\, d\mu_n = \int f\, d\mu$ *for all* $f \in \mathcal{L}$

The proof that i) follows from ii) involves the property of the open sets of being $\mathcal{L}$-open. In fact, given an open set $A$, $1_A = \sup u_n$, where $u_n = n \inf(\frac{1}{n}, d(x, X - A)) \in \mathcal{L}$, because $d(x, B) \in \mathcal{L}$ for every closed set $B$. By the same property it follows that $\mathcal{B}(X)$ coincides with $\mathcal{A}(\mathcal{L})$.

The completeness of $(\mathcal{M}^1, \delta)$, that is involved yet in the pointwise convergence of $(\mathcal{T}_N(\mu))_{N\in\mathbb{N}}$, depends on the relationship between measures and linear functionals of $\mathcal{C}^b(X)$.

Given a Cauchy sequence $(\mu_n)_{n\in\mathbb{N}}$ in $(\mathcal{M}^1, \delta)$, for any $f \in \mathcal{L}$, the real sequence $(\int f\, d\mu_n)_{n\in\mathbb{N}}$ too is a Cauchy sequence, and then, it converges to a limit that we denote by

$$L(f) = \lim_{n\to+\infty} \int f\, d\mu_n \tag{16}$$

$L : \mathcal{L} \to \mathbb{R}$, inherits from the integrals $\int f\, d\mu_n$ the property of being a positive linear form.

We then assign to the functions in $\mathcal{L}$ the role of elementary functions and to $L$ that of integral of elementary functions.

As well as the usual set of elementary functions, $\mathcal{L}$ is a Stone vector lattice that contains the constant functions.

We will prove that the positive linear form $L$ also satisfies the properties of an integral on a set of elementary functions, namely it is *monotone*, i.e. for any decreasing sequence $(f_n)_{n\in\mathbb{N}}$ in $\mathcal{L}$ with $\inf_{n\in\mathbb{N}} f_n = 0$,

$$\lim_{n\to+\infty} L(f_n) = 0.$$

When $X$ is locally compact, this property follows from Dini's theorem. In Lemma 3.3, we will prove that, in our case, the monotonicity property follows from the *uniform tightness* of the Cauchy sequences in $(\mathcal{M}^1, \delta)$, that is proved in Lemma 3.2.

We recall that, by the completeness and separability of $X$, any measure $\mu \in \mathcal{M}^1$ is *tight*, i.e. for every $\varepsilon > 0$, there exists a compact set $C_\varepsilon(\mu)$ such that $\mu(C_\varepsilon(\mu)) > 1 - \varepsilon$, and that a sequence $(\mu_n)_{n\in\mathbb{N}}$ in $\mathcal{M}^1$ is said to be uniformly tight, if for every $\varepsilon > 0$, there exists a compact set $C_\varepsilon$ such that $\mu_n(C_\varepsilon) > 1 - \varepsilon$ for all $\mu_n$.

Once proved that $L$ is monotone, we can apply, to the Stone vector lattice $\mathcal{L}$ and to the linear form $L$, the Daniell–Stone theorem:

**Theorem 3.1 (Daniell–Stone [1, Thm 7.1.4.])** *Let $\mathcal{V}$ be a Stone vector lattice of real functions on $X$ that contains the constant functions and let $L$ be a monotone* *positive linear form on $\mathcal{V}$. Then there exists exactly one measure $\mu$ on $\mathcal{A}(\mathcal{V})$, such that*

*(i) every $f \in \mathcal{V}$ is $\mu$-integrable;*
*(ii) $L(f) = \int f\, d\mu$ for all $f \in \mathcal{V}$.*

*Moreover $\mu(A) = inf\{\mu(G) : A \subseteq G,\ G$ subset $\mathcal{V}$-open of $X\}$.*

So we prove the existence of a measure $\mu \in \mathcal{M}^1$, such that

$$\int f\, d\mu = L(f) = \lim_{n\to+\infty} \int f\, d\mu_n \qquad \text{for all } f \in \mathcal{L}.$$

By Lemma 3.1, $(\mu_n)_{n\in\mathbb{N}}$ converges weakly to $\mu$.

Moreover, since $(\mu_n)_{n\in\mathbb{N}}$ is a Cauchy sequence in $(\mathcal{M}^1, \delta)$, the convergence of $(\int f\, d\mu_n)_{n\in\mathbb{N}}$ is uniform with respect to $f \in \mathcal{L}_1$ and then, the sequence $(\mu_n)_{n\in\mathbb{N}}$ converges, also in the metric $\delta$, to $\mu$.

We must then prove that the linear functional $L$, generated by the integrals of the measures of a Cauchy sequence $(\mu_n)_{n\in\mathbb{N}}$, is monotone.

**Lemma 3.2** *Let $(\mu_n)_{n\in\mathbb{N}}$ be a Cauchy sequence in $(\mathcal{M}^1, \delta)$, then it is uniformly tight.*

***Proof*** Since $X$ is separable there exists a countable set $\{q_i : i \in \mathbb{N}\}$ dense in $X$. For $\eta > 0$, we denote $B_i^{\eta} = \overline{B}(q_i, \eta)$.

For any $\varepsilon > 0$, let $\varepsilon_k = \frac{\varepsilon}{2^{k+1}}$. Choose $\overline{n}_k$ such that for any $n \geq \overline{n}_k$, $\delta(\mu_n, \mu_{\overline{n}_k}) < \frac{\varepsilon_k}{k}$ and $\tilde{N}_k$ such that $\mu_{\overline{n}_k}(\bigcup_{i=1}^{\tilde{N}_k} B_i^{\frac{1}{k}}) > 1 - \varepsilon_k$.

For $i = 1, 2, \ldots \tilde{N}_k$, set $\phi_i(p) = k(\frac{1}{k} - \min(\frac{1}{k}, d(p, B_i^{\frac{1}{k}}))$ and define

$$f(p) = \max\{\phi_i(p) : i = 1, 2, .., \tilde{N}_k\}.$$

It is easy to see that $0 \leq f(p) \leq 1$, $\mathrm{Lip}\,(f) \leq k$ and

$$f(p) = 1, \quad \text{if } p \in \bigcup_{i=1}^{\tilde{N}_k} B_i^{\frac{1}{k}}, \qquad f(p) = 0, \quad \text{if } p \notin \bigcup_{i=1}^{\tilde{N}_k} B(B_i^{\frac{1}{k}}, \frac{1}{k}).$$

Then

$$1 - \varepsilon_k < \mu_{\overline{n}}(\bigcup_{i=1}^{\tilde{N}_k} B_i^{\frac{1}{k}}) \leq \int f\, d\mu_{\overline{n}} = \int f\, d\mu_n + \int f\, d\mu_{\overline{n}} - \int f\, d\mu_n \leq$$

$$\int f\, d\mu_n + k\,\delta(\mu_n, \mu_{\overline{n}}) \leq \int f\, d\mu_n + k\frac{\varepsilon_k}{k} = \int f\, d\mu_n + \varepsilon_k$$

for all $n \geq \overline{n}$. Therefore, for these $n$,

$$1 - 2\varepsilon_k < \int f\, d\mu_n \leq \mu_n(\bigcup_{i=1}^{\tilde{N}_k} B(B_i^{\frac{1}{k}}, \frac{1}{k})) \leq \mu_n(\bigcup_{i=1}^{\tilde{N}_k} B_i^{\frac{2}{k}}),$$

since $B(B_i^{\frac{1}{k}}, \frac{1}{k}) \subset B_i^{\frac{2}{k}}$, for all $i$.

For any $n \in \mathbb{N}$, let $N(n,k)$ be such that

$$\mu_n(\bigcup_{i=1}^{N(n,k)} B_i^{\frac{2}{k}}) > 1 - 2\varepsilon_k$$

and let $\overline{N}_k = \max\{N(1,k), N(2,k), \ldots, N(\overline{n}-1,k), \tilde{N}_k\}$, then, for every $n \in \mathbb{N}$

$$\mu_n(\bigcup_{i=1}^{\overline{N}_k} B_i^{\frac{2}{k}}) > 1 - 2\varepsilon_k = 1 - \frac{\varepsilon}{2^k}$$

Let us set $C_k = \bigcup_{i=1}^{\overline{N}_k} B_i^{\frac{2}{k}}$. By proceeding as in [1, Thm. 7.3.3], we can prove, by induction, that $\mu_n(\bigcap_{k=1}^m C_k) > 1 - \varepsilon \sum_{k=1}^m \frac{1}{2^k}$ and, because of the completeness of $X$, that $C_\varepsilon = \bigcap_{k\in\mathbb{N}} C_k$ is the compact set that we looked for, indeed it satisfies

$$\mu_n(C_\varepsilon) > 1 - \varepsilon \qquad \text{for all } n \in \mathbb{N}.$$

□

**Lemma 3.3** *Let $(\mu_n)_{n\in\mathbb{N}}$ be a Cauchy sequence in $(\mathcal{M}^1, \delta)$, then the positive linear form $L : \mathcal{L} \to \mathbb{R}$, defined in (16), is monotone.*

***Proof*** Let $(f_i)_{i\in\mathbb{N}}$ be a sequence in $\mathcal{L}$ decreasing to 0.

Let us remark that $||f_i|| \leq ||f_1||$ for every $i \in \mathbb{N}$.

For any fixed $\varepsilon > 0$, let $C_\varepsilon$ be a compact set such that $\mu_n(C_\varepsilon) > 1 - \varepsilon$ for all $\mu_n$. Then, for any $n > 0$

$$\int f_i\, d\mu_n = \int_{C_\varepsilon} f_i\, d\mu_n + \int_{X-C_\varepsilon} f_i\, d\mu_n \leq$$

$$||f_i||_{|C_\varepsilon}\mu_n(C_\varepsilon) + ||f_1||\mu_n(X - C_\varepsilon) \leq$$

$$||f_i||_{|C_\varepsilon} + ||f_1||\,\varepsilon.$$

By passing to the limit with respect to $n$, we obtain $L(f_i) \leq ||f_i||_{|C_\varepsilon} + ||f_1||\,\varepsilon$. Since, by Dini's theorem, $||f_i||_{|C_\varepsilon}$ converges to 0, there exists $\overline{i}$ such that, for every

$i \geq \bar{i}$, $||f_i||_{C_\varepsilon} < \varepsilon$ and then $L(f_i) \leq (1 + ||f_1||)\,\varepsilon$. Therefore $L(f_i)$ converges to 0. □

By the previous Lemma, from Daniell–Stone theorem it follows that any Cauchy sequence in $(\mathcal{M}^1, \delta)$ converges to a measure in $\mathcal{M}^1$.

We have proved that

**Theorem 3.2** *$(\mathcal{M}^1, \delta)$ is a complete metric space.*

Now we define $\mathcal{T}$ and prove that it is a contraction on $(\mathcal{M}^1, \delta)$.

**Proposition 3.2** *For any $\mu \in \mathcal{M}^1$, the sequence $(\mathcal{T}_N(\mu))_{N\in\mathbb{N}}$ converges in the metric $\delta$ to a measure in $\mathcal{M}^1$, that we denote by $\mathcal{T}(\mu)$.*

***Proof*** Let $\mu \in \mathcal{M}^1$, $f \in \mathcal{L}_1$, and $q_0 \in X$ then, for any $N \in \mathbb{N}$ and $j > 0$

$$|\int f\, dT_N(\mu) - \int f\, dT_{(N+j)}(\mu)| =$$

$$|\int (f(q) - f(q_0))dT_N(\mu) - \int (f(q) - f(q_0))dT_{(N+j)}(\mu)| =$$

$$|\sum_{i=0}^{N} m_i^N \int (f\circ \mathrm{S}_i(p) - f\circ \mathrm{S}_i(\,\mathrm{S}_i^{-1}(q_0))d\mu - \sum_{i=0}^{N+j} m_i^{N+j} \int (f\circ \mathrm{S}_i(p) - f\circ \mathrm{S}_i(\mathrm{S}_i^{-1}(q_0))d\mu| \leq$$

$$\sum_{i=0}^{N}(m_i^N - m_i^{N+j}) \int (|f\circ \mathrm{S}_i(p) - f\circ \mathrm{S}_i(\,\mathrm{S}_i^{-1}(q_0))|d\mu$$

$$+ \sum_{i=N+1}^{N+j} m_i^{N+j} \int |f\circ \mathrm{S}_i(p) - f\circ\ \mathrm{S}_i(\mathrm{S}_i^{-1}(q_0))|d\mu \leq$$

$$(\sum_{i=0}^{N}(m_i^N - m_i^{N+j}) + \sum_{i=N+1}^{N+j} m_i^{N+j})\,\frac{1}{2}\ \mathrm{diam}\,(\,\mathrm{supp}\ \mu\,)$$

since $\mathrm{Lip}\,(f\circ \mathrm{S}_i) \leq \frac{1}{2}$.

By taking into account (15), from the previous inequalities it follows

$$|\int f\, dT_N(\mu) - \int f\, dT_{(N+j)}(\mu)| \leq (\sum_{i=N+1}^{N+j} m_i^{N+j})(\sum_{i=0}^{N} m_i^N + 1)\,\frac{1}{2}\ \mathrm{diam}\,(\,\mathrm{supp}\ \mu\,) \leq$$

$$\sum_{i=N+1}^{N+j} m_i^{N+j}\ \mathrm{diam}\,(\,\mathrm{supp}\ \mu\,) = \frac{\sum_{i=N+1}^{N+j} m_i}{\sum_{i=0}^{N+j} m_i}\ \mathrm{diam}\,(\,\mathrm{supp}\ \mu\,)$$

From the convergence of $\sum_{i=0}^{+\infty} m_i$ to 1, it follows that, for any fixed $\varepsilon > 0$, by choosing $\overline{N}$ sufficiently large, for any $N > \overline{N}$ and for any $j > 0$,

$$\frac{\sum_{i=N+1}^{N+j} m_i}{\sum_{i=0}^{N+j} m_i} \operatorname{diam}(\operatorname{supp} \mu) < \varepsilon. \tag{17}$$

Therefore, for any $\mu$,

$$\delta(T_N(\mu), T_{(N+j)}(\mu) \leq \varepsilon \text{ whenever } N > \overline{N}.$$

By the completeness of $(\mathcal{M}^1, \delta)$, the Cauchy sequence $(\mathcal{T}_N(\mu))_{N\in\mathbb{N}}$ converges to a measure that we denote by $\mathcal{T}(\mu)$. □

By proceeding as in [3, Thm 4.4.1.], we prove

**Lemma 3.4**

*(i) For any $N \geq 1$, $\mathcal{T}_N$ is a contraction of contraction factor $\frac{1}{2}$ on $(\mathcal{M}^1, \delta)$.*
*(ii) There exists exactly one $\mu \in \mathcal{M}^1$, that we denote by $\mu_N$, such that*

$$\mathcal{T}_N(\mu_N) = \mu_N.$$

***Proof***

(i) Let $\mu, \nu \in \mathcal{M}^1$, $f \in \mathcal{L}_1$, then

$$|\int f dT_N(\mu) - \int f dT_N(\nu)| =$$

$$|\sum_{i=0}^{N} m_i^N (\int f \circ \mathrm{S}_i \, d\mu - \int f \circ \mathrm{S}_i \, d\nu)| =$$

$$\frac{1}{2} |\sum_{i=0}^{N} m_i^N (\int 2f \circ \mathrm{S}_i \, d\mu - \int 2f \circ \mathrm{S}_i \, d\nu)| \leq$$

$$\frac{1}{2} \sum_{i=0}^{N} m_i^N \delta(\mu, \nu) = \frac{1}{2} \delta(\mu, \nu),$$

since $\operatorname{Lip}(2f \circ \mathrm{S}_i) \leq 1$. Then $\delta(\mathcal{T}_N(\mu), \mathcal{T}_N(\nu)) \leq \frac{1}{2} \delta(\mu, \nu)$.

(ii) follows from (i) and Theorem 3.2.

□

**Theorem 3.3**

*(i)* $\mathcal{T}$ *is a contraction of contraction factor* $\frac{1}{2}$ *on* $(\mathcal{M}^1\ \delta)$.
*(ii) There exists exactly one* $\mu \in \mathcal{M}^1$, *that we denote by* $\mu_\infty$, *such that*

$$\mathcal{T}(\mu) = \mu.$$

*(iii)*

$$\text{supp}\,\mu_\infty = K.$$

*(iv) The sequence* $(\mu_N)_{N\in\mathbb{N}}$ *converges to* $\mu_\infty$.

***Proof*** Let $\mu, \nu \in \mathcal{M}^1$, with $\text{supp}\,\mu \bigcup \text{supp}\,\nu \subset B(K, r)$. For $\varepsilon > 0$, choose $\overline{N}$ so that (17) holds, with diam supp $\mu$ replaced by $2r$. Since it does not depend on $\mu$ or $\nu$, for any $N > \overline{N}$

$$\delta(\mathcal{T}(\mu), \mathcal{T}(\nu)) \leq \delta(\mathcal{T}(\mu), \mathcal{T}_N(\mu)) + \delta(\mathcal{T}_N(\mu), (\mathcal{T}_N(\nu)) + \delta(\mathcal{T}_N(\nu), \mathcal{T}(\nu)) \leq$$

$$2\varepsilon + \delta(\mathcal{T}_N(\mu), \mathcal{T}_N(\nu)).$$

Then, by the previous Lemma and arbitrariness of $\varepsilon$, $\delta(\mathcal{T}(\mu), \mathcal{T}(\nu)) \leq \frac{1}{2}\,\delta(\mu, \nu)$.

(ii) follows from (i).

(iii) As in the finite dimensional case, the claim follows from the invariance of $\text{supp}\,\mu_\infty$ with respect to $\mathscr{S}$.

(iv)

$$\delta(\mu_\infty, \mu_N) = \delta(T(\mu_\infty), T_N(\mu_N)) \leq$$

$$\delta(\mathcal{T}(\mu_\infty), \mathcal{T}_N(\mu_\infty)) + \delta(\mathcal{T}_N(\mu_\infty), \mathcal{T}_N(\mu_N)) \leq \delta(\mathcal{T}(\mu_\infty), \mathcal{T}_N(\mu_\infty)) + \frac{1}{2}\delta(\mu_\infty, \mu_N).$$

Then

$$\delta(\mu_\infty, \mu_N) \leq 2\,\delta(\mathcal{T}(\mu_\infty), \mathcal{T}_N(\mu_\infty))$$

and

$$\lim_{N\to+\infty} \delta(\mu_\infty, \mu_N) \leq \lim_{N\to+\infty} 2\,\delta(\mathcal{T}(\mu_\infty), \mathcal{T}_N(\mu_\infty)) = 0.$$

□

We have proved that there exists exactly one measure invariant with respect to $\mathscr{S}$ and $\mathcal{W}$. Actually, by ii) of the previous Theorem, Lemma 3.1, and Proposition 3.1, $\mu_\infty$ is the only measure that satisfies

$$\mu_\infty = \sum_{k=0}^{+\infty} m_k \mu_\infty \circ \text{S}_k^{-1}.$$

## References

1. Bauer, H.: Probability Theory and Elements of Measure Theory. Holt, Rinehart and Winston Inc., New York (1972)
2. Beherends, E., Kadets, V.M.: Metric spaces with the small ball property. Stud. Math. **148**, 275–287 (2001)
3. Goodey, P.R.: Generalized Hausdorff dimension. Mathematika **17**, 324–327 (1970)
4. Hutchinson, J.E.: Fractals and self-similarity. Indiana Univ. Math. J. **30**, 713–747 (1981)
5. Kravchenko, A.S.: Completeness of the space of separable measures in the Kantorovich-Rubinstein metric. Sib. Math. J. **47**(1), 68–76 (2006)
6. Parthasarathy, K.R.: Probability Measures in Metric Spaces. Academic Press, London (1967)
7. Zitkovic, G.: Theory of Probability. Lecture Notes. The University of Texas at Austin, Department of Mathematics (2013). http://web.ma.utexas.edu

# On the External Approximation of Sobolev Spaces by $M$-Convergence

**Umberto Mosco and Maria Agostina Vivaldi**

**Abstract** In this paper we deal with the external approximation of the Sobolev spaces $H^1(\Omega)$ and $H_0^1(\Omega)$ in coordinate Sobolev spaces via $M$-convergence.

**Keywords** $M$-convergence · External approximation · Sobolev spaces

## 1 Introduction

Several applications of the $M$-convergence to convergence and approximation of Lebesgue and Sobolev spaces have been given in [10–12], in particular, in Lemma 6.4, Lemma 6.5, and Corollary 6.1 of [11], and approximations of Sobolev spaces in a general setting are described in Chapter IV of [6] and Chapter 1, Chapter 5 of [1]. The aim of this paper is to provide a self-contained presentation of the external approximation of the Sobolev space $H^1(\Omega)$ in the product Sobolev space $X = H^1(\Omega)_1 \times H^1(\Omega)_2$ based on the $M$-convergence of subspaces in Hilbert spaces, as illustrated in [10].

External approximations of Sobolev spaces have recently proved to be useful in dealing with the so-called SOC theory, as shown recently in [13] and [14].

Before describing the plan of this paper in more detail, let us briefly comment on the reason why external approximations in coordinate product Sobolev spaces—in particular Theorem 3.1 of the present paper—are relevant in our approach to the SOC theory presented in Mosco and Vivaldi [14]. We recall that SOC models appeared first in the physics literature at the turn of eighties and nineties in papers related to avalanches and sand pile discrete particle models. The basic

U. Mosco
Department of Mathematics, Worcester Polytechnic Institute, Worcester, MA, USA
e-mail: mosco@wpi.edu

M. A. Vivaldi (✉)
Dip. Di Scienze di base ed Applicate per l'ingegneria, Sapienza Università di Roma, Roma, Italy
e-mail: maria.vivaldi@sbai.uniroma1.it



M. R. Lancia, A. Rozanova-Pierrat (eds.), *Fractals in Engineering: Theoretical Aspects and Numerical Approximations*, SEMA SIMAI Springer Series 8,
https://doi.org/10.1007/978-3-030-61803-2_6

idea underlying these papers was that certain nonlinear dissipative systems drive themselves autonomously in a finite time to a critical state, which is the attractor of the dynamics. Later, a PDE approach to sand piles was developed by Barbu in [2] and [3] in the framework of multivalued nonlinear diffusion equations. The main contribution of our paper [14] is to provide a rigorous mathematical proof of the self-organized criticality SOC property in a fully discrete setting, where the limit brings from finitely many particles to infinitely many particles. This discrete theory, more closely related to the physical models, can be formulated in the settings of the coordinate Sobolev spaces $X$ and $X_0$ and of their $M$-convergence to the usual differential Sobolev spaces $H^1$ and $H^1_0$. In view of this important application of the $M$-convergence theory, in the present paper we provide a self-contained description of the main features of $M$-convergence—which involve both weak and strong convergences of sequences in Hilbert spaces—and of the application to the approximation of Sobolev spaces, as already mentioned in the Introduction.

The plan of this paper is the following. In Sect. 2 we give some preliminary results about discrete finite grids in the plane and construct the coordinate Sobolev spaces. In Sect. 3 we give our main results. The proofs will be given in Sects. 4 and 5, and some comments on fractal examples in Sect. 6.

## 2 Sobolev Space Approximations

By $\Omega = \Omega_L$, where $L > 0$, we denote throughout this paper the open square domain of $\mathbb{R}^2$

$$\Omega = (-L, L) \times (-L, L).$$

On $\Omega$, we consider the following coordinate Sobolev spaces:

$$H^1(\Omega)_i = \{v \in L^2(\Omega) : \frac{\partial v}{\partial x_i} \in L^2(\Omega)\}, \quad i = 1, 2,$$

which are Hilbert spaces with the inner product

$$(u, v)_{H^1(\Omega)_i} = \int_\Omega (uv + \frac{\partial v}{\partial x_i}\frac{\partial u}{\partial x_i})dx, \ i = 1, 2,$$

and the product space

$$X \equiv H^1(\Omega)_1 \times H^1(\Omega)_2$$

of all pairs $u = (u_1, u_2)$, $u_1 \in H^1(\Omega)_1$, $u_2 \in H^1(\Omega)_2$, which is a Hilbert space with the product norm

$$||u||_X = \left(||u_1||^2_{L^2(\Omega)} + ||\frac{\partial u_1}{\partial x_1}||^2_{L^2(\Omega)} + ||u_2||^2_{L^2(\Omega)} + ||\frac{\partial u_2}{\partial x_2}||^2_{L^2(\Omega)}\right)^{\frac{1}{2}}. \tag{2.1}$$

The aim of this section is to describe suitable finite dimensional approximations of the spaces introduced before.

To this end, we start by constructing in $\Omega$ coordinate discrete grids $G_m$, $m \in \mathbb{N}$, of increasing cardinality as $m \to +\infty$. For every $m$, we set

$$h = h_m = \frac{2L}{2^m} \tag{2.2}$$

and define the discrete grid $G_m$ in $\Omega$, the boundary grid $\partial G_m$ in $\partial\Omega$, and the grid $\overline{G_m}$ in $\overline{\Omega}$—all of mesh size $h$—as follows:

$$G_m := \left\{\left(q_1\frac{2L}{2^m}, q_2\frac{2L}{2^m}\right) : (q_1, q_2) \in \mathbb{Z}^2, |q_1| \le 2^{m-1} - 1, |q_2| \le 2^{m-1} - 1\right\}$$

$$\partial G_m := \left\{\left(q_1\frac{2L}{2^m}, q_2\frac{2L}{2^m}\right) : (q_1, q_2) \in \mathbb{Z}^2, |q_1| = 2^{m-1}, |q_2| \le 2^{m-1} - 1\right\}$$

$$\cup \left\{\left(q_1\frac{2L}{2^m}, q_2\frac{2L}{2^m}\right) : (q_1, q_2) \in \mathbb{Z}^2, |q_1| = 2^{m-1} - 1, |q_2| = 2^{m-1}\right\}$$

$$\cup \left\{\left(q_1\frac{2L}{2^m}, q_2\frac{2L}{2^m}\right) : (q_1, q_2) \in \mathbb{Z}^2, |q_1| = 2^{m-1}, |q_2| = 2^{m-1}\right\}$$

$$\overline{G_m} := G_m \cup \partial G_m .$$

$G_m$ is composed of $(2^m - 1) \times (2^m - 1)$ points in $\Omega$, while $\partial G_m$ is composed of $2^m 4$ points on $\partial\Omega$. Moreover, the grid $G_m$ decomposes the square $\Omega$ into coordinate square elements of side $h$. The pictures below show the grids $G_2$ and $G_3$ in the unit square (Fig. 1).

From now on, we denote the generic point of the grid $\overline{G_m}$ by

$$(q_1\frac{2L}{2^m}, q_2\frac{2L}{2^m}) = (jh, kh).$$

At every point $(jh, kh) \in \overline{G_m}$, we define the *cross-like region* $I \cup J$, where

$$\begin{cases} I := \{x = (x_1, x_2) : (j-1)h \le x_1 \le (j+1)h,\ (k - \frac{1}{2})h \le x_2 \le (k + \frac{1}{2})h\} \\ J := \{x = (x_1, x_2) : (j - \frac{1}{2})h \le x_1 \le (j + \frac{1}{2})h,\ (k-1)h \le x_2 \le (k+1)h\}. \end{cases} \tag{2.3}$$

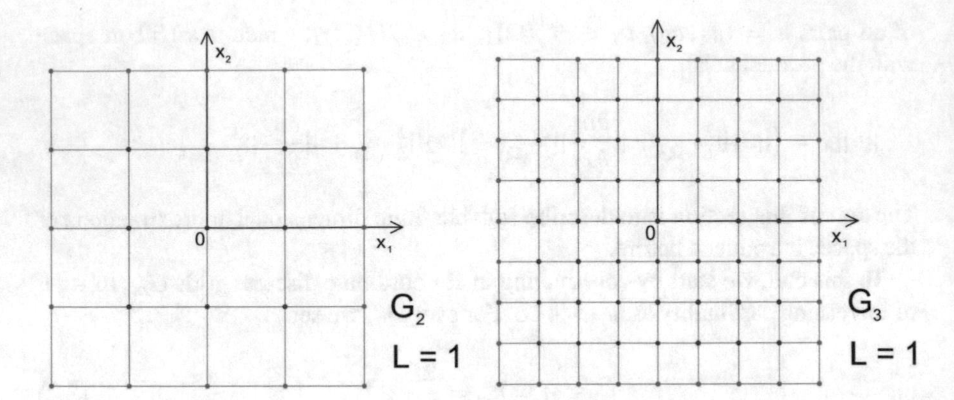


**Fig. 1** $G_2$ and $G_3$

The central square element of this cross-like region, that is, the set

$$I \cap J = \{x = (x_1, x_2) : (j - \frac{1}{2})h \leq x_1 \leq (j + \frac{1}{2})h, \ (k - \frac{1}{2})h \leq x_2 \leq (k + \frac{1}{2})h\},$$

will be called the *central square element* of the cross.

The function

$$\theta_{jk}(x) = \chi(\frac{x_1}{h} - j)\chi(\frac{x_2}{h} - k)$$

where $\chi(s)$, $s \in \mathbb{R}$, denotes the characteristic function of the real interval $[\frac{-1}{2}, \frac{1}{2}) \subset \mathbb{R}$, is supported on the central square element $I \cap J$ of the cross, as it is easily checked.

We then introduce the pair of horizontal and vertical *tent functions*

$$\varphi^1_{jk}(x) = \varphi(\frac{x_1}{h} - j)\,\chi(\frac{x_2}{h} - k), \quad \varphi^2_{jk}(x) = \chi(\frac{x_1}{h} - j)\,\varphi(\frac{x_2}{h} - k),$$

where

$$\varphi(s) = \chi * \chi(s)$$

is obtained by taking the convolution of the characteristic function $\chi$ of the interval $[\frac{-1}{2}, \frac{1}{2}) \subset \mathbb{R}$ with itself. It is easy to check that

$$supp\, \varphi^1_{jk}(x) = I, \qquad supp\, \varphi^2_{jk}(x) = J\,.$$

The picture of the tent functions and their supports is below (Fig. 2).

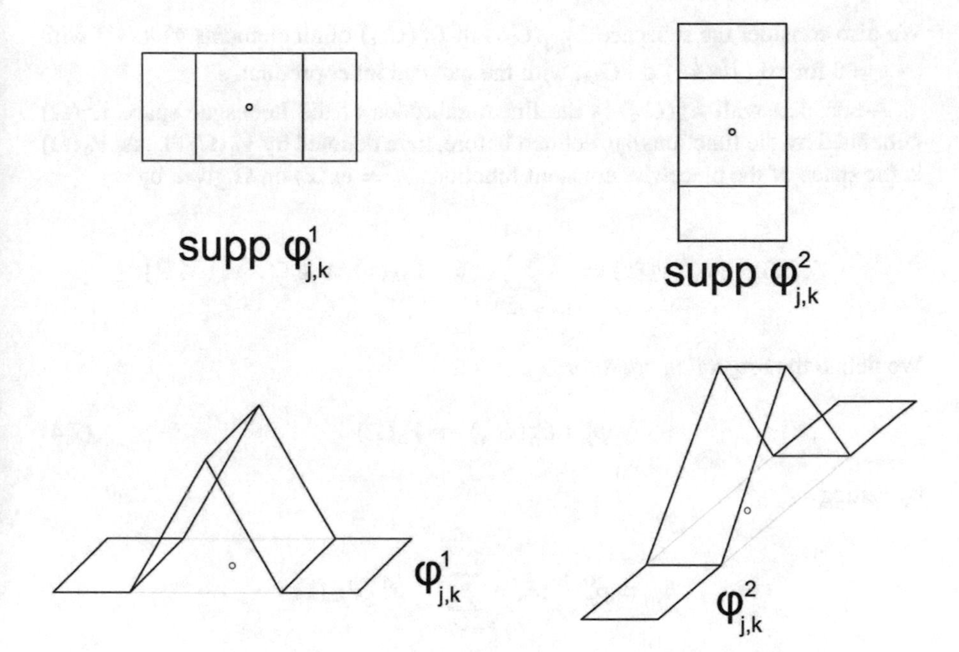


**Fig. 2** Tent functions and their supports

If we fix $m \in \mathbb{N}$, then $h$ also remains fixed, namely, $h = h_m = 2^{-m}2L$. Then, in our previous notation

$$\left(q_1\frac{2L}{2^m}, q_2\frac{2L}{2^m}\right) = (jh, kh)$$

for the points of the grid $\overline{G}_m$, for each $m$ we consider the space of all elements $v^h$ obtained by attaching a real value $v^h_{j,k}$ to every point (i.e. node) of the grid $\overline{G}_m$. We denote such a space by $V^h$, that is,

$$V^h = \{v^h : v_{j,k},\ \ v_{j,k} \in \mathbb{R},\ \ j, k = -2^{m-1}, \ldots\ldots., 2^{m-1}\}.$$

Then we introduce the (finite dimensional) Hilbert space

$$L^2_h(G_m)$$

of all elements $v^h \in V^h$ with the inner product of two elements $v^h$ and $u^h$ of $V^h$ defined by

$$(v^h, u^h)_h := h^2 \sum_{j,k=-2^{m-1}}^{2^{m-1}} v^h_{j,k}\, u^h_{j,k}.$$

We also consider the subspace $L^2_{h,0}(G_m)$ of $L^2_h(G_m)$ of all elements $v^h \in V^h$ with $v^h_{j,k} = 0$ for all $(jh, kh) \in \partial G_m$, with the induced inner product.

Associated with $L^2_h(G_m)$ is the linear subspace of the Lebesgue space $L^2(\Omega)$ generated by the functions $\theta_{jk}$ defined before, here denoted by $V_h(\Omega)$. Thus, $V_h(\Omega)$ is the space of the piecewise constant functions $v_h = v_h(x)$ on $\Omega$ given by

$$V_h(\Omega) = \{v_h : v_h(x) = \sum_{j,k=-2^{m-1}}^{2^{m-1}} v_{j,k}\, \theta_{jk}(x),\ x \in \Omega\,,\ v_{j,k} \in \mathbb{R}\}\,.$$

We define the *projection operator*

$$p^0_h : L^2_h(G_m) \to V_h(\Omega) \tag{2.4}$$

by setting

$$v_h = p^0_h v^h := \sum_{j,k=-2^{m-1}}^{2^{m-1}} v^h_{j,k}\, \theta_{jk}(x)$$

and the *restriction operator*

$$r_h : L^2(\Omega) \to L^2_h(G_m) \tag{2.5}$$

by setting

$$r_h v = v^h =: \{v^h_{j,k}\},$$

where

$$v^h_{j,k} = h^{-2} \int_\Omega v(x)\, \theta_{jk}(x)\, dx$$

is the *average* of $v(x)$ over the support of $\theta_{jk}$. We recall that the support of $\theta_{jk}$ is the central square element $I \cap J$ of the cross. For the operators $p^0_h$ and $r_h$, the following results hold.

**Proposition 2.1**

$$\frac{1}{2}||v^h||_{L^2(G_m)} \le ||p^0_h v^h||_{L^2(\Omega)} \le ||v^h||_{L^2(G_m)},\ \ \forall v_h \in L^2_h(G_m) \tag{2.6}$$

$$||r_h v||_{L^2_h(G_m)} \le ||v||_{L^2(\Omega)},\ \ \forall v \in L^2(\Omega) \tag{2.7}$$

$$||v - p^0_h\, r_h\, v||_{L^2(\Omega)} \to 0\ \ as\ \ h \to 0,\ \ \forall v \in L^2(\Omega). \tag{2.8}$$

***Proof*** We first notice that as the supports do not overlap then

$$|p_h^0 v_h(x)|^2 = |\sum_{j,k=-2^{m-1}}^{2^{m-1}} v_{j,k}^h \, \theta_{jk}(x)|^2 = \sum_{j,k=-2^{m-1}}^{2^{m-1}} |v_{j,k}^h|^2 \theta_{jk}(x).$$

Thus we deduce (2.6) as the integral in $\Omega$ of the function $\theta_{jk}$ is equal to $h^2$ if the node $(jh, kh)$ belongs to $G_m$, it is equal to $h^2/2$ if the node $(jh, kh)$ belongs to $\partial G_m$, and it is different from a vertex of $\Omega$, and it is equal to $h^2/4$ if the node $(jh, kh)$ is a vertex of $\Omega$. By Cauchy–Schwarz inequality,

$$h^{-4} | \int_\Omega v(x) \, \theta_{jk}(x) \, dx|^2 \le h^{-4} \Big( \int_\Omega \theta_{jk}(x) \, dx \Big) \Big( \int_\Omega |v(x)|^2 \, \theta_{jk}(x) \, dx \Big)$$

and

$$|r_h v|_h^2 \le \sum_{j,k=-2^{m-1}}^{2^{m-1}} \int_\Omega |v(x)|^2 \, \theta_{jk}(x) \, dx \ \le \int_\Omega |v(x)|^2 \, dx,$$

that is (2.7). To prove (2.8), we can assume that the function $v$ is continuous because the space of the continuous functions on $\bar{\Omega}$ is dense in $L^2(\Omega)$.

Therefore,

$$\begin{aligned} |v(x) - p_h^0 r_h v(x)|^2 &= | \sum_{j,k=-2^{m-1}}^{2^{m-1}} (v(x) - v_{j,k}^h) \, \theta_{jk}(x)|^2 \\ &= \sum_{j,k=-2^{m-1}}^{2^{m-1}} (v(x) - v_{j,k}^h)^2 \, \theta_{jk}(x) = A + B, \end{aligned}$$

where we have split the last sum into two parts $A$ and $B$ that are, respectively, the sum on the indices $j, k$ such that the node $(jh, kh)$ belongs to $G_m$, and the sum on the *boundary* indices that are the indices $j, k$ such that the node $(jh, kh)$ belongs to $\partial G_m$.

We first prove that

$$\int_\Omega A \, dx = \int_\Omega \sum_{j,k=-2^{m-1}+1}^{2^{m-1}-1} (v(x) - v_{j,k}^h)^2 \, \theta_{jk}(x) dx \le \omega(v, \sqrt{2}h), \tag{2.9}$$

where $\omega(v, \sqrt{2}h) = \sup_{|y| \le \sqrt{2}h} \int_\Omega |v(x-y) - v(x)|^2 dx$.

We recall that for every function $v \in L^2(\Omega)$ we have $\omega(v, \sqrt{2}h) \to 0$ as $h \to 0$.

To show (2.9), we note that

$$\sum_{j,k=-2^{m-1}+1}^{2^{m-1}-1} (v(x)-v^h_{j,k})^2\,\theta_{jk}(x) = \sum_{j,k=-2^{m-1}+1}^{2^{m-1}-1} \Big(\frac{1}{h^2}\int_\Omega (v(x)-v(t))\theta_{jk}(t)d\,t\Big)^2\,\theta_{jk}(x).$$

Now if $x \in supp\ \theta_{jk}$ and $t \in supp\ \theta_{jk}$, then $|y| = |x-t| \le \sqrt{2}h$ and

$$|v(x)-v(t)|^2 \le \sup_{|y|\le\sqrt{2}h} |v(x-y)-v(x)|^2.$$

Hence (by Cauchy–Schwarz inequality),

$$\begin{cases} \int_\Omega \sum_{j,k=-2^{m-1}+1}^{2^{m-1}-1} (\frac{1}{h^2}\int_\Omega (v(x)-v(t))\theta_{jk}(t)d\,t)^2\,\theta_{jk}(x)d\,x \le \\ \int_\Omega \sum_{j,k=-2^{m-1}+1}^{2^{m-1}-1} \frac{h^2}{h^4}\int_\Omega (v(x)-v(t))^2\theta_{jk}(t)d\,t\ \ \theta_{jk}(x)d\,x \\ \le \sup_{|y|\le\sqrt{2}h}\int_\Omega |v(x-y)-v(x)|^2 dx. \end{cases} \tag{2.10}$$

Now we consider the remaining part $B$ that is the sum of four *similar* terms $B_i$, $i = 1, 2, 3, 4$ that are the sum on the indices $j, k$ such that the node $(jh, kh)$ belongs to the four sides of the square $\Omega$. We estimate $B_1$ as the other contributions can be evaluated analogously.

$$B_1 = \sum_{j=-2^{m-1}}^{2^{m-1}} (v(x)-v^h_{j,2^{m-1}})^2\,\theta_{j2^{m-1}}(x)$$

$$\begin{cases} \int_\Omega \sum_{j=-2^{m-1}}^{2^{m-1}} (v(x)-v^h_{j,2^{m-1}})^2\,\theta_{j2^{m-1}}(x)d\,x \le \\ 2\Big(\int_\Omega \sum_{j=-2^{m-1}}^{2^{m-1}} v(x)^2\,\theta_{j2^{m-1}}(x)d\,x + \int_\Omega \sum_{j=-2^{m-1}}^{2^{m-1}} (v^h_{j,2^{m-1}})^2\,\theta_{j2^{m-1}}(x)d\,x\Big). \end{cases} \tag{2.11}$$

We note that for every function $v \in L^2(\Omega)$ the first integral in the right hand side of inequality (2.10) tends to 0 as $h \to 0$ and the second one can be bounded in terms of the first one. More precisely, we have (by Cauchy–Schwarz inequality)

$$\int_\Omega \sum_{j=-2^{m-1}}^{2^{m-1}} (v^h_{j,2^{m-1}})^2\,\theta_{j2^{m-1}}(x)d\,x$$

$$= \int_\Omega \sum_{j=-2^{m-1}}^{2^{m-1}} \Big(\frac{1}{h^2}\int_\Omega v(t)\theta_{j2^{m-1}}(t)d\,t\Big)^2\,\theta_{j2^{m-1}}(x)d\,x \le$$

$$\int_{\Omega} \sum_{j=-2^{m-1}}^{2^{m-1}} \frac{h^2}{h^4} \int_{\Omega} v(t)^2 \theta_{j2^{m-1}}(t) d\,t \;\; \theta_{j2^{m-1}}(x) d\,x =$$

$$\sum_{j=-2^{m-1}}^{2^{m-1}} \frac{h^2}{h^4} \int_{\Omega} v(t)^2 \theta_{j2^{m-1}}(t) d\,t \;\; \int_{\Omega} \theta_{j2^{m-1}}(x) d\,x \leq \sum_{j=-2^{m-1}}^{2^{m-1}} \frac{h^4}{2h^4} \int_{\Omega} v(t)^2 \theta_{j2^{m-1}}(t) d\,t.$$

The statement in (2.8) is showed, and then the proof Proposition 2.1 is achieved. □

With tent functions $\varphi^1_{jk}(x)$ and $\varphi^2_{jk}(x)$ introduced before, we construct suitable approximating linear subspaces

$$X_h \subset X$$

of the product space $X = H^1(\Omega)_1 \times H^1(\Omega)_2$

$$X_h = \{(v^1(x)_{|\Omega}, v^2(x)_{|\Omega})\},$$

where

$$v^1(x) = \sum_{jk=-2^{m-1}}^{2^{m-1}} v_{j,k}\, \varphi^1_{jk}(x), \qquad v^2(x) = \sum_{jk=-2^{m-1}}^{2^{m-1}} v_{j,k}\, \varphi^2_{jk}(x)\,, \tag{2.12}$$

the coefficients $v_{j,k}$ being variable in $\mathbb{R}$. By taking coefficients $v_{j,k} = 0$ in the previous sums whenever the point $(jh, kh)$ is a boundary point $(jh, kh) \in \partial\Omega_m$ of $\Omega$, we get the subspace $X_{h,0} \subset X_h$

$$X_{h,0} = \{(v^1(x), v^2(x)) \in X_h : v^1(x) = v^2(x) = 0 \;\; \forall x \in \partial\Omega\}\,. \tag{2.13}$$

The Sobolev space $H^1(\Omega)$ can be identified with *diagonal* of $X$. We also introduce the *projection operator*

$$p^1_h : L^2_h(G_m) \to X_h \tag{2.14}$$

by setting

$$p^1_h(v_h) = (v^1(x), v^2(x)),$$

where

$$v^1(x) = \sum_{j,k=-2^{m-1}}^{2^{m-1}} v^h_{j,k}\, \varphi^1_{jk}(x), \qquad v^2(x) = \sum_{j,k=-2^{m-1}}^{2^{m-1}} v^h_{j,k}\, \varphi^2_{jk}(x).$$

## 3 The $M$-Convergence Result

In this section, as said in the introduction, we state our main result, Theorem 3.1 below. As said in the Introduction, this result refers to the external approximation of the Sobolev spaces $H^1(\Omega)$ and $H^1_0(\Omega)$ in the product Sobolev space $X = H^1(\Omega)_1 \times H^1(\Omega)_2$ in terms of the $M$-convergence of subspaces in a Hilbert space.

We recall the definition of $M$-convergence of subspaces of a Hilbert space introduced in [10].

**Definition 1** A sequence of subspaces $M_h$ of a Hilbert space $X$ is said to $M$-converge to a subspace $M$ of $X$ as $h \to 0$ if the two properties below hold:

(i) for every $u \in M$, there exists $u_h \in M_h$ strongly converging to $u$ in $X$;
(ii) for every sequence $v_h$ of $X$ and every subsequence $v_{h_k}$ of $v_h$ weakly converging to $v$ in $X$ as $k \to +\infty$, we have $v \in M$.

Our main result is the following theorem.

**Theorem 3.1** *The following properties hold:*

(j) *the space $X_h$ $M$-converges to the space $H^1(\Omega)$ in $X$ as $h \to 0$;*
(jj) *the space $X_{h,0}$ $M$-converges to the space $H^1_0(\Omega)$ in $X$ as $h \to 0$.*

In view of the definition of $M$-convergence given above, Theorem 3.1 is an immediate consequence of Lemma 1 and Lemma 2 below.

**Lemma 1** *We have:*

(b) *for every sequence $v_h \in X_h$ and every subsequence $v_{h_k}$ of $v_h$ weakly converging to $u$ in $X$ as $k \to +\infty$, $u$ belongs to $H^1(\Omega)$;*
(c) *for every sequence $v_h \in X_{h,0}$ and every subsequence $v_{h_k}$ of $v_h$ weakly converging to $u$ in $X$ as $k \to +\infty$, $u$ belongs to $H^1_0(\Omega)$.*

**Lemma 2** *We have:*

(a) *for every $u \in H^1(\Omega)$, the elements $u_h = p^1_h r_h u$ belong to $X_h$ and converge strongly to $u$ in $X$;*
(aa) *if $u \in H^1_0(\Omega)$, then there exist elements in $X_{h,0}$ strongly converging to $u$ in $X$.*

In the statement (a) of Lemma 2, the operator $p^1_h$ is defined in (2.14) and the operator $r_h$ is defined in (2.5).

To show that the two lemmas imply Theorem 3.1, it suffices to observe that property (j) follows from (b) and (a), and property (jj) from (c) and (aa).

The proof of Lemma 1 will be given in Sect. 4, and the proof of Lemma 2 will be given in Sect. 5.

# 4 Proof of Lemma 1

*Proof of property (b) of Lemma 1*

The proof of property (b) is based on a preliminary estimate. The estimate involves the function

$$v_h(x) = (v_h^1(x), v_h^2(x)) = \Big( \sum_{j,k=-2^{m-1}}^{2^{m-1}} v_{j,k}^h\, \varphi_{jk}^1(x), \sum_{j,k=-2^{m-1}}^{2^{m-1}} v_{j,k}^h\, \varphi_{jk}^2(x) \Big)$$

and the function

$$w_h(x) = \sum_{j,k=-2^{m-1}}^{2^{m-1}} v_{j,k}^h\, \theta_{jk}(x) .$$

We shall prove that for every $h$ we have

$$\|w_h\|_{L^2(\Omega)}^2 \le C\, \|v_h^1\|_{L^2(\Omega)}^2 , \tag{4.1}$$

where the constant $C$ does not depend on $h$.

By the expressions of $v_h^1$, as the supports of the functions $\chi(\frac{x_2}{h} - k)$ do not overlap,

$$\int_\Omega \Big( \sum_{j,k=-2^{m-1}}^{2^{m-1}} v_{j,k}^h\, \varphi_{j,k}^1(x)\Big)^2 dx = \int_\Omega \sum_{k=-2^{m-1}}^{2^{m-1}} \Big( \sum_{j=-2^{m-1}}^{2^{m-1}} v_{j,k}^h\, \varphi(\frac{x_1}{h} - j)\Big)^2 \chi(\frac{x_2}{h} - k) dx_1 dx_2. \tag{4.2}$$

We evaluate the double integral over $\Omega$ as two successive simple integrals from $-L$ to $L$, first in the variable $x_2$ and then in the variable $x_1$. The integral between $-L$ and $L$ in the variable $x_2$ of the function $\chi(\frac{x_2}{h} - k)$ is equal to $h$ if $k$ is different from $+2^{m-1}$ and from $-2^{m-1}$, and it is equal to $h/2$ if $k$ is equal to $+2^{m-1}$ or equal to $-2^{m-1}$. This gives

$$\begin{cases} \int_\Omega (\sum_{j,k=-2^{m-1}}^{2^{m-1}} v_{j,k}^h\, \varphi_{j,k}^1(x))^2 d\,x = \\ h \sum_{k=-2^{m-1}+1}^{2^{m-1}-1} \int_{-L}^{L} (\sum_{j=-2^{m-1}}^{2^{m-1}} v_{j,k}^h\, \varphi(\frac{x_1}{h} - j))^2 d\,x_1 + \\ \frac{h}{2} \int_{-L}^{L} (\sum_{j=-2^{m-1}}^{2^{m-1}} v_{j,-2^{m-1}}^h\, \varphi(\frac{x_1}{h} - j))^2 d\,x_1 + \\ \frac{h}{2} \int_{-L}^{L} (\sum_{j=-2^{m-1}}^{2^{m-1}} v_{j,2^{m-1}}^h\, \varphi(\frac{x_1}{h} - j))^2 d\,x_1 . \end{cases} \tag{4.3}$$

Now we evaluate the first integral at the right hand side of (4.3).

The integral in $dx_1$ can be computed as the sum of the integrals over the intervals $[jh, (j+1)h]$, with $j$ running from $-2^{m-1}$ to $+2^{m-1}-1$. On each interval $[jh, (j+$

$1)h]$ (with $j \neq 2^{m-1}$), we obtain only two non-zero values. In fact, in $[jh, (j+1)h]$ we have $\varphi(\frac{x_1}{h} - j) = -\frac{x_1}{h} + j + 1$ and $\varphi(\frac{x_1}{h} - (j+1)) = +\frac{x_1}{h} - (j+1) + 1$. Therefore, the function to be integrated is $((v^h_{j+1,k} - v^h_{j,k})(\frac{x_1}{h} - j) + v_{j,k})^2$. The integral of this function is equal to

$$\frac{h}{3}\big((v^h_{j+1,k})^2 + (v^h_{j,k})^2 + v^h_{j+1,k} v^h_{j,k}\big)\,.$$

Here are the details of this calculation

$$\int_{jh}^{(j+1)h} ((v^h_{j+1,k} - v^h_{j,k})(\frac{x_1}{h} - j) + v_{j,k})^2\, dx_1 =$$

$$= \int_{jh}^{(j+1)h} \{(v^h_{j+1,k} - v^h_{j,k})^2 (\frac{x_1}{h} - j)^2 + (v^h_{j,k})^2 +$$

$$+2v^h_{j,k}(v^h_{j+1,k} - v^h_{j,k})(\frac{x_1}{h} - j)\}dx_1 =$$

$$= (v^h_{j+1,k} - v^h_{j,k})^2 \frac{h}{3}[(j+1-j)^3 - (j-j)^3] + (v^h_{j,k})^2 h +$$

$$+2v^h_{j,k}(v^h_{j+1,k} - v^h_{j,k})\frac{h}{2}[(j+1-j)^2 - (j-j)^2] =$$

$$= \frac{h}{3}[(v^h_{j+1,k})^2 + (v^h_{j,k})^2 - 2v^h_{j+1,k}v^h_{j,k} + 3(v^h_{j,k})^2 + 3v^h_{j,k}v^h_{j+1,k} - 3(v^h_{j,k})^2] =$$

$$= \frac{h}{3}[(v^h_{j,k})^2 + (v^h_{j+1,k})^2 + v^h_{j,k}v^h_{j+1,k}]\,.$$

In this way we obtain

$$\begin{cases} \int_{-L}^{L} (\sum_{j=-2^{m-1}}^{2^{m-1}} v^h_{j,k}\,\varphi(\frac{x_1}{h} - j))^2 d\,x_1 = \\ \sum_{j=-2^{m-1}}^{2^{m-1}-1} \int_{jh}^{(j+1)h} ((v^h_{j+1,k} - v^h_{j,k})(\frac{x_1}{h} - j) + v_j)^2 d\,x_1 = \\ \frac{h}{3}\sum_{j=-2^{m-1}}^{2^{m-1}-1} \big((v^h_{j+1,k})^2 + (v^h_{j,k})^2 + v^h_{j+1,k}v^h_{j,k}\big). \end{cases} \tag{4.4}$$

At this point, by using the elementary inequality $A^2 + B^2 + AB \geq (1/2)(A^2 + B^2)$, we finally get

$$\int_{-L}^{L} (\sum_{j=-2^{m-1}}^{2^{m-1}} v^h_{j,k}\,\varphi(\frac{x_1}{h} - j))^2 d\,x_1 \geq \frac{h}{6} \sum_{j=-2^{m-1}}^{2^{m-1}-1} \big((v^h_{j+1,k})^2 + (v^h_{j,k})^2\big).$$

Analogous estimates hold for the second and last integrals in the right hand side of (4.3). Combining together (4.3) and (4.4), we then obtain

$$\begin{cases} \int_\Omega (\sum_{j,k=-2^{m-1}}^{2^{m-1}} v_{j,k}^h \varphi_{j,k}^1(x))^2 dx \geq \\ \sum_{k=-2^{m-1}}^{2^{m-1}} \frac{h^2}{12} \sum_{j=-2^{m-1}}^{2^{m-1}-1} \left((v_{j+1,k}^h)^2 + (v_{j,k}^h)^2\right). \end{cases} \tag{4.5}$$

As seen before, the left hand side of this inequality is equal to $\|v_h^1\|^2_{L^2(\Omega)}$.

Moreover we note that $\theta_{jk}(x) = \theta_{jk}^2(x)$ for every $j$ and $k$, the supports of the functions $\theta_{jk}$ do not overlap, and $\int_\Omega \theta_{j,k}(x)\,dx = h^2$ (in the *interior* nodes), and then, we obtain

$$\begin{cases} \|w_h\|^2_{L^2(\Omega)} = h^2\{\sum_{j,k=-2^{m-1}+1}^{2^{m-1}-1} (v_{j,k}^h)^2 + \frac{1}{2}\sum_{k=-2^{m-1}+1}^{2^{m-1}-1} (v_{2^{m-1},k}^h)^2 + \\ +\frac{1}{2}\sum_{k=-2^{m-1}+1}^{2^{m-1}-1} (v_{-2^{m-1},k}^h)^2 + \frac{1}{2}\sum_{j=-2^{m-1}+1}^{2^{m-1}-1} (v_{j,-2^{m-1}}^h)^2 \\ +\frac{1}{2}\sum_{j=-2^{m-1}+1}^{2^{m-1}-1} (v_{j,2^{m-1}}^h)^2 + \frac{1}{4}\Big((v_{-2^{m-1},-2^{m-1}}^h)^2 + (v_{2^{m-1},2^{m-1}}^h)^2 + \\ (v_{-2^{m-1},+2^{m-1}}^h)^2 + (v_{2^{m-1},-2^{m-1}}^h)^2\Big)\}. \end{cases} \tag{4.6}$$

Hence,

$$\|w_h\|^2_{L^2(\Omega)} \leq h^2 \sum_{j,k=-2^{m-1}}^{2^{m-1}} (v_{j,k}^h)^2, \tag{4.7}$$

and estimate (4.1) follows from formulas (4.5) and (4.7).

This concludes the proof of our preliminary estimate (4.1), and from this point on, the proof of (b) proceeds as follows.

By the assumption in condition (b), the sequence $v_{h_k} \in X_{h_k}$ weakly converges in $X = H^1(\Omega)_1 \times H^1(\Omega)_2$; therefore, $v_{h_k}^1$ weakly converges in $L^2(\Omega)$ to a function $v^*$ and $v_{h_k}^2$ weakly converges in $L^2(\Omega)$ to a function $v^{**}$; moreover, the sequence $\frac{\partial v_{h_k}^1}{\partial x_1}$ weakly converges in $L^2(\Omega)$ to a function $v^{*1}$ and the sequence $\frac{\partial v_{h_k}^2}{\partial x_2}$ weakly converges in $L^2(\Omega)$ to a function $v^{*2}$.

From now on, to simplify the notation, we denote all such subsequences generically by a single suffix $h$. In the notation just set, the sequence $v_h^1$ is bounded in $L^2(\Omega)$. By taking into account estimate (4.1), we deduce that the sequence $w_h$, by possibly extracting a further subsequence, weakly converges in $L^2(\Omega)$ to a function $w$.

We claim that

$$w = v^* = v^{**}. \tag{4.8}$$

The strategy to prove this claim is to prove separately, first that $v^* = w$ and then that $v^{**} = w$.

To show that

$$v^* = w, \tag{4.9}$$

we start by proving that

$$\int_\Omega (w - v^*) dx = 0. \tag{4.10}$$

To prove (4.10), we develop the integral over $\Omega$ as follows:

$$\begin{cases} \int_\Omega (v_h^1 - w_h) dx = -\int_\Omega \sum_{j,k=-2^{m-1}}^{2^{m-1}} v_{j,k}^h (\theta_{j,k}(x) - \varphi_{j,k}^1(x)) dx = \\ -h \sum_{k=-2^{m-1}+1}^{2^{m-1}-1} \sum_{j=-2^{m-1}}^{2^{m-1}-1} \int_{jh}^{(j+1)h} \sum_{j=-2^{m-1}}^{2^{m-1}} v_{j,k}^h (\chi(\frac{x_1}{h} - j) - \varphi(\frac{x_1}{h} - j)) dx_1 + \\ -\frac{h}{2} \sum_{j=-2^{m-1}}^{2^{m-1}-1} \int_{jh}^{(j+1)h} \sum_{j=-2^{m-1}}^{2^{m-1}} v_{j,-2^{m-1}}^h (\chi(\frac{x_1}{h} - j) - \varphi(\frac{x_1}{h} - j)) dx_1 + \\ -\frac{h}{2} \sum_{j=-2^{m-1}}^{2^{m-1}-1} \int_{jh}^{(j+1)h} \sum_{j=-2^{m-1}}^{2^{m-1}} v_{j,2^{m-1}}^h (\chi(\frac{x_1}{h} - j) - \varphi(\frac{x_1}{h} - j)) dx_1. \end{cases} \tag{4.11}$$

The internal sum over $j$ inside the first term of the sum at the right hand side of (4.11) vanishes: in fact, by taking into account the definition of the functions $\varphi$ and $\chi$, we have

$$\begin{cases} \sum_{j=-2^{m-1}}^{2^{m-1}-1} \int_{jh}^{(j+1)h} \sum_{j=-2^{m-1}}^{2^{m-1}} v_{j,k}^h (\varphi(\frac{x_1}{h} - j) - \chi(\frac{x_1}{h} - j)) dx_1 = \\ \sum_{j=-2^{m-1}}^{2^{m-1}-1} \int_{jh}^{(j+1/2)h} \sum_{j=-2^{m-1}}^{2^{m-1}} v_{j,k}^h (\varphi(\frac{x_1}{h} - j) - \chi(\frac{x_1}{h} - j)) dx_1 + \\ \sum_{j=-2^{m-1}}^{2^{m-1}-1} \int_{(j+1/2)h}^{(j+1)h} \sum_{j=-2^{m-1}}^{2^{m-1}} v_{j,k}^h (\varphi(\frac{x_1}{h} - j) - \chi(\frac{x_1}{h} - j)) dx_1 = \\ \sum_{j=-2^{m-1}}^{2^{m-1}-1} \int_{jh}^{(j+1/2)h} (v_{j+1,k}^h - v_{j,k}^h)(\frac{x_1}{h} - j)) dx_1 + \\ \sum_{j=-2^{m-1}}^{2^{m-1}-1} \int_{(j+1/2)h}^{(j+1)h} ((v_{j+1,k}^h - v_{j,k}^h)(\frac{x_1}{h} - j) + v_{j,k}^h - v_{j+1,k}^h) dx_1 = 0, \end{cases} \tag{4.12}$$

and analogous estimates hold for the second and last terms at the right hand side of (4.11), showing that these terms too vanish.

In a similar way we prove that

$$\int_R (v_h^1 - w_h) dx = 0 \tag{4.13}$$

for every choice of integers $M_1, M_2, N_1, N_2 \in \mathbb{Z}$ with $\max\{|N_1|, |N_2|, |M_1|, |M_2|\} \le 2^{m-1}$, $N_1 < M_1, N_2 < M_2$ and any rectangle $R = R(M_1, M_2, N_1, N_2) = \bigcup_{j=N_1}^{M_1-1} \bigcup_{k=N_2}^{M_2-1} [jh, (j+1)h] \times [kh, (k+1)h]$. Moreover for any set $E$ and any positive $\varepsilon$, we can choose (for $h$ sufficiently small) a finite union $\Re$ of rectangles

$R(M_1, M_2, N_1, N_2)$ of the type mentioned before such that $\Re \subset E$ and the measure of $E \setminus \Re$ is less than $\varepsilon$. Then,

$$|\int_E (v^* - w) d\,x| = |\lim_{h\to 0} \int_E (v_h^1 - w_h) d\,x| = |\lim_{h\to 0} \int_{E\setminus\Re} (v_h^1 - w_h) d\,x$$

$$+ \int_{\Re} (v_h^1 - w_h) d\,x| \le \sqrt{\varepsilon} ||v_h^1 - w_h||_{L^2(\Omega)}. \tag{4.14}$$

The proof of (4.9) is complete.

We recall that $v^{**}$ is the weak limit in $L^2(\Omega)$ of the sequence $v_h^2$. We can then repeat the previous calculations by replacing $v_h^1$ by $v_h^2$, and we get

$$w = v^{**}.$$

This concludes the proof of (4.8.)

In particular, we have shown that the two components $v_h^1$ and $v_h^2$ of the vector $v_h$ have the same weak limit. From now on, we denote this limit by $u$ and now we show that

$$u \in H^1(\Omega). \tag{4.15}$$

In order to prove (4.15), we show that the function $v^{*1}$, which is the weak limit of the sequence $\frac{\partial v_h^1}{\partial x_1}$, is equal to $\frac{\partial u}{\partial x_1}$ and the function $v^{*2}$, the weak limit of the sequence $\frac{\partial v_h^2}{\partial x_1}$, is equal to $\frac{\partial u}{\partial x_2}$. In fact, let $\phi(x)$ be an infinitely differentiable function with support on $\Omega$, then

$$\int_\Omega \frac{\partial v_h^1}{\partial x_1} \phi\, d\,x = -\int_\Omega v_h^1 \frac{\partial \phi}{\partial x_1}\, d\,x \text{ and } \int_\Omega \frac{\partial v_h^2}{\partial x_2} \phi\, d\,x = -\int_\Omega v_h^2 \frac{\partial \phi}{\partial x_2}\, dx\,.$$

Passing to the limit as $h \to 0$, we then obtain

$$\int_\Omega v^{*1} \phi\, d\,x = -\int_\Omega u \frac{\partial \phi}{\partial x_1}\, d\,x \text{ and } \int_\Omega v^{*2} \phi d\,x = -\int_\Omega u \frac{\partial \phi}{\partial x_2}\, d\,x.$$

Thus, $v^{*1} = \frac{\partial u}{\partial x_1}$ and $v^{*2} = \frac{\partial u}{\partial x_2}$ and, in particular, $u \in H^1(\Omega)$.

The proof of condition (b) of Lemma 1 is now completed.

We move now to the *Proof of property (c) of Lemma 1*.

We must show that if $v_h$ belongs to the subspace $X_{h,0}$, then the trace of $u$ on $\partial\Omega$ is zero. The proof is as follows. First, we note that the first components $v_h^1$ of the vector $v_h \in X_h$ have trace on the vertical sides of domain $\Omega$.

More precisely (for any fixed $h$) the trace $\gamma_{h1} = \gamma_{h1}(-L)$ of $v_h^1$ on the side $\{-L\} \times [-L, L]$ is the strong limit as $\varepsilon \to 0$ in the space $L^2(-L, L)$ of the function

$$\frac{1}{\varepsilon} \int_{-L}^{-L+\varepsilon} v_h^1(x_1, x_2) d x_1,$$

and the following estimate holds

$$||\gamma_{h1}||^2_{L^2(-L,L)} \leq C\{||v_h^1||^2_{L^2(\Omega)} + ||\frac{\partial v_h^1}{\partial x_1}||^2_{L^2(\Omega)}, \} \tag{4.16}$$

where the constant $C$ does not depend on $h$.

Analogously the trace of $v_h^1$ on the side $\{L\} \times [-L, L]$ is the strong limit as $\varepsilon \to 0$ in the space $L^2(-L, L)$ of the function $\frac{1}{\varepsilon} \int_{L-\varepsilon}^{L} v_h^1(x_1, x_2) d x_1$, the trace of $v_h^2$ on the side $[-L, L] \times \{-L\}$ is the strong limit as $\varepsilon \to 0$ in the space $L^2(-L, L)$ of the function $\frac{1}{\varepsilon} \int_{-L+\varepsilon}^{-L} v_h^2(x_1, x_2) d x_2$, and the trace of $v_h^2$ on the side$[-L, L] \times \{L\}$ is the strong limit as $\varepsilon \to 0$ in the space $L^2(-L, L)$ of the function $\frac{1}{\varepsilon} \int_{L-\varepsilon}^{L} v_h^2(x_1, x_2) d x_2$, and estimates analogous to (4.16) hold.

In particular if the sequence $v_h$ is bounded in the space $X$, then the sequences of traces of the functions $v_h^1$ on the vertical sides of domain $\Omega$ and the sequences of traces of the functions $v_h^2$ on the horizontal sides of domain $\Omega$ are bounded in the space $L^2(-L, L)$. Hence (up to pass to a subsequence) the sequence of the functions $\gamma_{h1}$ weakly converges in $L^2(-L, L)$ to a function $\gamma_1$.

Moreover let $\phi(x)$ be an infinitely differentiable function vanishing in a neighborhood of the segment $\{L\} \times [-L, L]$, then we have

$$\int_\Omega \frac{\partial v_h^1}{\partial x_1} \phi \, d x = - \int_\Omega v_h^1 \frac{\partial \phi}{\partial x_1} d x + \int_{-L}^{L} \gamma_{h1} \phi \, d x_2.$$

Passing to the limit (as $h \to 0$), we obtain

$$\int_\Omega \frac{\partial u}{\partial x_1} \phi \, d x = - \int_\Omega u \frac{\partial \phi}{\partial x_1} d x + \int_{-L}^{L} \gamma_1 \phi \, d x_2,$$

and hence, the function $\gamma_1$ is the trace of the function $u$ on the side $\{-L\} \times [-L, L]$.

Now if the vectors $v_h$ belong to the subspace $X_{h,0}$, then the traces of $v_h^1$ on the vertical sides of domain $\Omega$ vanish and, analogously, the traces of $v_h^2$ on the horizontal sides of domain $\Omega$ vanish (see formula (2.13)), and we conclude that the function $u$ vanishes on the boundary of the domain $\Omega$.

This completes the proof of property (c) of Lemma 1. Thus, the proof of Lemma 1 has been achieved.

# 5 Proof of Lemma 2

The proof is given in various steps.

*We prove Property (a) of Lemma 2.*

According to the definition of the norm in the space $X$ (see (2.1)), we have to prove that

$$\begin{cases} ||v - v_h^1||_{L^2(\Omega)} \to 0 \text{ as } h \to 0, \ \forall v \in H^1(\Omega), \\ ||v - v_h^2||_{L^2(\Omega)} \to 0 \text{ as } h \to 0, \ \forall v \in H^1(\Omega) \end{cases} \tag{5.1}$$

and

$$\begin{cases} ||\frac{\partial v}{\partial x_1} - \frac{\partial v_h^1}{\partial x_1}||_{L^2(\Omega)} \to 0 \text{ as } h \to 0, \ \forall v \in H^1(\Omega), \\ ||\frac{\partial v}{\partial x_2} - \frac{\partial v_h^2}{\partial x_2}||_{L^2(\Omega)} \to 0 \text{ as } h \to 0, \ \forall v \in H^1(\Omega). \end{cases} \tag{5.2}$$

We recall that

$$p_h^1 r_h v = (v_h^1(x), v_h^2(x)) \text{ where } v_{j,k} = (r_h v)_{j,k} \text{ and } v_h^i(x) = \sum_{j,k=-2^{m-1}}^{2^{m-1}} v_{j,k}\, \varphi_{jk}^i(x) \ \ i = 1, 2.$$

We first prove the first statement in (5.2). We can assume that the function $v$ belongs to the space $C^1(\bar{\Omega})$ because this space is dense in $H^1(\Omega)$.

An easy calculation shows that

$$\frac{\partial v_h^1}{\partial x_1} = \sum_{j,k=-2^{m-1}}^{2^{m-1}} \frac{v_{j+1,k} - v_{j,k}}{h} \chi_j^*(x_1)\chi_k(x_2),$$

where $\chi_k(x_2) = \chi(\frac{x_2}{h} - k)$ is the characteristic function of the interval $[(k - 1/2)h, (k+1/2)h)$ and $\chi_j^*(x_1)$ is the characteristic function of the interval $[jh, (j+1)h)$.

From now on in this proof, in order to simplify the notation, we set

$$\frac{\partial v}{\partial x_1} = w. \tag{5.3}$$

By repeating the proof of the statement (2.8) in Proposition 2.1, we can deduce that

$$||w - p_h^{0,*}\, r_h\, w||_{L^2(\Omega)} \to 0 \text{ as } h \to 0, \tag{5.4}$$

where we set $p_h^{0,*}\, r_h\, w = \sum_{j,k=-2^{m-1}}^{2^{m-1}} w_{j,k}^*\, \chi_j^*(x_1)\chi_k(x_2)$, and $w_{j,k}^* = h^{-2}\int_\Omega w(x)\,\chi_j^*(x_1)\chi_k(x_2)\,dx$ (compare the definition of $p_h^0 r_h w$ with the definition of $p_h^{0,*} r_h w$ and the definition of $w_{j,k}$ with the definition of $w_{j,k}^*$).

Having in mind (5.4) in order to show (5.2), we have to prove that

$$||p_h^{0,*}\, r_h\, w - \sum_{j,k=-2^{m-1}}^{2^{m-1}} \frac{v_{j+1,k}-v_{j,k}}{h}\, \chi_j^*(x_1)\chi_k(x_2)||_{L^2(\Omega)} \to 0 \text{ as } h \to 0. \tag{5.5}$$

As already noticed, we have

$$\Big(\sum_{j,k=-2^{m-1}}^{2^{m-1}} (w_{j,k}^* - \frac{v_{j+1,k}-v_{j,k}}{h})\, \chi_j^*(x_1)\chi_k(x_2)\Big)^2 =$$

$$\sum_{j,k=-2^{m-1}}^{2^{m-1}} (w_{j,k}^* - \frac{v_{j+1,k}-v_{j,k}}{h})^2 \chi_j^*(x_1)\chi_k(x_2) = A + B,$$

where we have split the last sum into two parts $A$ and $B$ that are, respectively, the sum on the indices $j, k$ such that the node $(jh, kh)$ belongs to $G_m$, and $j \le 2^{m-1}-2$ and the sum on the remaining indices. We first prove that

$$\begin{cases} \int_\Omega A\,dx = \int_\Omega \sum_{j=-2^{m-1}+1}^{2^{m-1}-2} \sum_{k=-2^{m-1}+1}^{2^{m-1}-1} (w_{j,k}^* - \frac{v_{j+1,k}-v_{j,k}}{h})^2 \chi_j^*(x_1)\chi_k(x_2)\,dx \le \\ C\omega(w, 2\sqrt{2}h), \end{cases} \tag{5.6}$$

where

$$\omega(w, 2\sqrt{2}h) = \sup_{|y|\le 2\sqrt{2}h} \int_\Omega |w(x-y) - w(x)|^2 dx$$

and $C$ is independent of $h$.

We have

$$(w_{j,k}^* - \frac{v_{j+1,k}-v_{j,k}}{h})^2 \le 2\Big((w_{j,k}^* - w(x))^2 + (w(x) - \frac{v_{j+1,k}-v_{j,k}}{h})^2\Big),$$

and we will estimate separately the integrals of the two terms on the right hand side.

By the definition of the coefficients $v_{j+1,k}$ and $v_{j,k}$, we have (for any node $(jh, kh)$ belonging to $G_m$, and such that $j \leq 2^{m-1} - 2$)

$$w(x)-\frac{v_{j+1,k}-v_{j,k}}{h}=\frac{1}{h^2}\int_{(j+1/2)h}^{(j+3/2)h}\int_{(k-1/2)h}^{(k+1/2)h}(w(x)-\frac{v(y)-v(y_1-h,y_2)}{h})dy_1dy_2=$$

$$\frac{1}{h^3}\int_{(j+1/2)h}^{(j+3/2)h}\int_{(k-1/2)h}^{(k+1/2)h}\int_{y_1-h}^{y_1}(w(x)-w(s,y_2))ds\,dy_1dy_2\leq$$

$$\frac{\sqrt{h}}{h^3}\int_{(j+1/2)h}^{(j+3/2)h}\int_{(k-1/2)h}^{(k+1/2)h}\left(\int_{(j-1/2)h}^{(j+3/2)h}(w(x)-w(s,y_2))^2ds\right)^{\frac{1}{2}}dy_1dy_2\leq$$

$$\frac{h^3}{h^3}\left(\sup_{|t|\leq 2\sqrt{2}h}|w(x)-w(x-t)|^2\right)^{\frac{1}{2}}.$$

Here $y = (y_1, y_2)$, and we have used (5.3), the Cauchy–Schwarz inequality, and the fact that if $x \in supp\ \chi_j^*(x_1)\chi_k(x_2)$ and $y \in supp\ \theta_{j+1\ k}$ then $|t| = |(x_1 - s, x_2 - y_2)| \leq 2\sqrt{2}h$. Moreover, as the *supports* of the functions $\chi_j^*(x_1)\chi_k(x_2)$ do not overlap, we have

$$\int_\Omega \sum_{j=-2^{m-1}+1}^{2^{m-1}-2}\ \sum_{k=-2^{m-1}+1}^{2^{m-1}-1}(w(x)-\frac{v_{j+1,k}-v_{j,k}}{h})^2\chi_j^*(x_1)\chi_k(x_2)dx\leq$$

$$\int_\Omega \sup_{|t|\leq 2\sqrt{2}h}|w(x)-w(x-t)|^2\sum_{j=-2^{m-1}+1}^{2^{m-1}-2}\ \sum_{k=-2^{m-1}+1}^{2^{m-1}-1}\chi_j^*(x_1)\chi_k(x_2)dx\leq$$

$$\int_\Omega \sup_{|t|\leq 2\sqrt{2}h}|w(x)-w(x-t)|^2dx=\omega(w,2\sqrt{2}h).$$

Analogously by the definition of the coefficients $w^*_{j,k}$, we have (for any node $(jh, kh)$ belonging to $G_m$ )

$$w^*_{j,k}-w(x)=\frac{1}{h^2}\int_{jh}^{(j+1)h}\int_{(k-1/2)h}^{(k+1/2)h}(w(y)-w(x))dy_1dy_2\leq,$$

$$\frac{h}{h^2}\left(\int_{jh}^{(j+1)h}\int_{(k-1/2)h}^{(k+1/2)h}(w(x)-w(y))^2dy_1dy_2\right)^{\frac{1}{2}}\leq,$$

$$\frac{h^2}{h^2}\left(\sup_{|t|\leq\sqrt{2}h}|w(x)-w(x-t)|^2\right)^{\frac{1}{2}}.$$

Here $y = (y_1, y_2)$ and we have used (5.3), the Cauchy–Schwarz inequality, and the fact that if $x \in supp\ \chi_j^*(x_1)\chi_k(x_2)$ and $y \in supp\ \chi_j^*(x_1)\chi_k(x_2)$, then $|t| = |(x_1 - y_1, x_2 - y_2)| \le \sqrt{2}h$.

Hence as the *supports* of the functions $\chi_j^*(x_1)\chi_k(x_2)$ do not overlap, we have

$$\int_\Omega \sum_{j=-2^{m-1}+1}^{2^{m-1}-2} \sum_{k=-2^{m-1}+1}^{2^{m-1}-1} (w^*_{j,k} - w(x))^2 \chi_j^*(x_1)\chi_k(x_2) d\,x \le$$

$$\int_\Omega \sup_{|t|\le 2\sqrt{2}h} |w(x) - w(x-t)|^2 \sum_{j=-2^{m-1}+1}^{2^{m-1}-2} \sum_{k=-2^{m-1}+1}^{2^{m-1}-1} \chi_j^*(x_1)\chi_k(x_2) d\,x \le$$

$$\int_\Omega \sup_{|t|\le\sqrt{2}h} |w(x) - w(x-t)|^2 d\,x = \omega(w, \sqrt{2}h).$$

The proof of estimate (5.6) is now completed, and we have to consider the remaining part $B$ that is the sum of five *similar* terms $B_i$, $i = 1, 2, 3, 4, 5$ that are the sum on the indices $j, k$ such that the node $(jh, kh)$ belongs to the four sides of the square $\Omega$ and the sum on $k$ for $j = 2^{m-1} - 1$. We estimate $B_1$ as the other contributions can be evaluated analogously.

$$B_1 = \sum_{j=-2^{m-1}}^{2^{m-1}} (w^*_{j,2^{m-1}} - \frac{v_{j+1,2^{m-1}} - v_{j,2^{m-1}}}{h})^2\ \chi_j^*(x_1)\chi_{2^{m-1}}(x_2),$$

$$\begin{cases} \int_\Omega \sum_{j=-2^{m-1}}^{2^{m-1}} (w^*_{j,2^{m-1}} - \frac{v_{j+1,2^{m-1}} - v_{j,2^{m-1}}}{h})^2\ \chi_j^*(x_1)\chi_{2^{m-1}}(x_2) d\,x \le \\ 2\Big( \int_\Omega \sum_{j=-2^{m-1}}^{2^{m-1}} (w^*_{j,2^{m-1}})^2\ \chi_j^*(x_1)\chi_{2^{m-1}}(x_2) d\,x + \\ \int_\Omega \sum_{j=-2^{m-1}}^{2^{m-1}} (\frac{v_{j+1,2^{m-1}} - v_{j,2^{m-1}}}{h})^2\ \chi_j^*(x_1)\chi_{2^{m-1}}(x_2) d\,x \Big). \end{cases} \tag{5.7}$$

We note that for every function $w \in L^2(\Omega)$ both the integrals in the right hand side of inequality (5.7) tend to 0 as $h \to 0$.

More precisely we have (by the Cauchy–Schwarz inequality)

$$\int_\Omega \sum_{j=-2^{m-1}}^{2^{m-1}} (w^*_{j,2^{m-1}})^2\ \chi_j^*(x_1)\chi_{2^{m-1}}(x_2) d\,x =$$

$$\int_\Omega \sum_{j=-2^{m-1}}^{2^{m-1}} \Big(\frac{1}{h^2} \int_{jh}^{(j+1)h} \int_{(k-1/2)h}^{(k+1/2)h} w(y)\, d\,y_1 d\,y_2\Big)^2 \chi_j^*(x_1)\chi_{2^{m-1}}(x_2) d\,x \le$$

$$\int_\Omega \sum_{j=-2^{m-1}}^{2^{m-1}} \frac{h^2}{h^4} \int_\Omega w(y)^2 \chi_j^*(y_1)\chi_{2^{m-1}}(y_2) d\,y\ \ \chi_j^*(x_1)\chi_{2^{m-1}}(x_2) d\,x =$$

$$\sum_{j=-2^{m-1}}^{2^{m-1}} \frac{h^2}{h^4} \int_\Omega w(y)^2 \chi_j^*(y_1)\chi_{2^{m-1}}(y_2) dy \int_\Omega \theta_{j2^{m-1}}(x) d\,x$$

$$\leq \sum_{j=-2^{m-1}}^{2^{m-1}} \frac{h^4}{2h^4} \int_\Omega w(y)^2 \theta_{j2^{m-1}}(y) d\,y.$$

Analogously,

$$\int_\Omega \sum_{j=-2^{m-1}}^{2^{m-1}} (\frac{v_{j+1,2^{m-1}} - v_{j,2^{m-1}}}{h})^2 \chi_j^*(x_1)\chi_{2^{m-1}}(x_2) d\,x =$$

$$\int_\Omega \sum_{j=-2^{m-1}}^{2^{m-1}} \Big(\frac{1}{h^3} \int_{(j+1/2)h}^{(j+3/2)h} \int_{(2^{m-1}-1/2)h}^{(2^{m-1}+1/2)h} \int_{y_1-h}^{y_1} w(s, y_2) d\,s\,d\,y_1 d\,y_2\Big)^2 \chi_j^*(x_1)\chi_{2^{m-1}}(x_2) d\,x \leq$$

$$\int_\Omega \sum_{j=-2^{m-1}}^{2^{m-1}} \frac{h^3}{h^6} \int_\Omega \int_{(j-1/2)h}^{(j+3/2)h} (w(s, y_2))^2 \chi_{j+1}(y_1)\chi_{2^{m-1}}(y_2)\, d\,sd\,y\ \ \chi_j^*(x_1)\chi_{2^{m-1}}(x_2) d\,x \leq$$

$$\sum_{j=-2^{m-1}}^{2^{m-1}} \frac{h^4}{h^6} \int_{(j-1/2)h}^{(j+3/2)h} \int_{(2^{m-1}-1/2)h}^{(2^{m-1}+1/2)h} ((w(s, y_2))^2\, d\,s\,d\,y_2 \int_\Omega \chi_j^*(x_1)\chi_{2^{m-1}}(x_2) d\,x$$

$$\leq \sum_{j=-2^{m-1}}^{2^{m-1}} \frac{h^6}{h^6} \int_{\Omega_{j,2^{m-1}}} (w(s, y_2))^2 d\,sd\,y_2,$$

where $\Omega_{j,2^{m-1}} = \Omega \bigcap \Big( supp\ \chi_j \chi_{2^{m-1}} \bigcup supp\ \chi_{j+1}\chi_{2^{m-1}} \Big)$, and we set $w = 0$ outside of $\Omega$.

As for any function $w \in L^2(\Omega)$, the integral $\omega(w, ah)$ vanishes as $h \to 0$ ($a > 0$ and independent of $h$) as well as the integral $\int_E w^2 d\,x$ vanishes as the Lebesgue measure of $E$ vanishes, then the statement in (5.5) is showed. By using (5.4) and (5.5), the proof of the first statement in (5.2) (see (5.3)) is also completed. We skip the proof of the second statement in (5.2) because we can easily repeat the previous arguments and we focus our attention in proving (5.1).

We begin by establishing the following estimate: there exists a positive constant $C$ (independent of $h$) such that

$$||v_h^1||_{L^2(\Omega)} \leq C, \tag{5.8}$$

where $v_h^1$ is the first component of the vector $p_h^1 r_h v$ and $v \in H^1(\Omega)$ that is

$$v_h^1 = \sum_{j,k=-2^{m-1}}^{2^{m-1}} v_{j,k}^h \, \varphi_{jk}^1.$$

As in the proof of condition (b) of Lemma 1 (see in particular (4.2), (4.3), and the related calculations in Sect. 4), we obtain

$$||v_h^1||_{L^2(\Omega)}^2 \leq \sum_{k=-2^{m-1}}^{2^{m-1}} \sum_{j-2^{m-1}}^{2^{m-1}-1} 2h^2\big((v_{j+1,k}^h)^2 + (v_{j,k}^h)^2\big). \tag{5.9}$$

By construction, $v_{j,k}^h$ is equal to $(r_h v)_{j,k}$, and we use estimate (2.7) of Proposition 2.1 (see the definition of the inner product in the space $L_h^2(G_m)$) to complete the proof of (5.8).

Now we use estimate (5.8) to show the first statement in (5.1).

From (5.8), we can get (up to extraction of subsequences) that

$$v_h^1 \rightharpoonup u^* \text{ in } L^2(\Omega) \text{ (weakly)}. \tag{5.10}$$

In the first part of the proof of condition (b) of Lemma 1(see Sect. 4), we have proved that if

$$v_h^1 = \sum_{j,k=-2^{m-1}}^{2^{m-1}} v_{j,k}^h \, \varphi_{jk}$$

converges in $L^2(\Omega)$ weakly to a function $v^*$, then also the sequence

$$\sum_{j,k=-2^{m-1}}^{2^{m-1}} v_{j,k}^h \, \theta_{jk}$$

converges in $L^2(\Omega)$ weakly to the *same limit*, that is, it converges in $L^2(\Omega)$ weakly to $v^*$.

Therefore, from (5.10), it follows that

$$p_h^0 r_h v = \sum_{j,k=-2^{m-1}}^{2^{m-1}} v_{j,k}^h \theta_{jk} \rightharpoonup u^*$$

(where $v_{j,k}^h = (r_h v)_{j,k}$).

By (2.8) of Proposition 2.1, we have that (for any $v \in L^2(\Omega)$ and by choosing $v_{j,k}^h = (r_h v)_{j,k}$)

$$p_h^0 r_h v = \sum_{j,k=-2^{m-1}}^{2^{m-1}} v_{j,k}^h \theta_{jk} \to v$$

in $L^2(\Omega)$ (strongly), therefore by the uniqueness of the limit

$$u^* = v \tag{5.11}$$

more than that *all subsequences*—hence the whole sequence—converge to the same limit $u^* = v$, and thus, we may replace (5.10) with

$$v_h^1 \rightharpoonup v \text{ in } L^2(\Omega) \text{ (weakly)}. \tag{5.12}$$

To show the strong convergence of $v_h^1 \to v$ in $L^2(\Omega)$, we will prove that

$$\limsup ||v_h^1||_{L^2(\Omega)} \le ||v||_{L^2(\Omega)} \,. \tag{5.13}$$

To prove the estimate (5.13), we come back to the inequalities in Sect. 4: more precisely we derive from the proof of condition (b) in Lemma 1 (see (4.2), (4.3), and (4.4)), where $v_{j,k}^h = (r_h v)_{j,k}$,

$$\begin{cases} ||v_h^1||_{L^2(\Omega)}^2 = \frac{h^2}{3} \sum_{k=-2^{m-1}+1}^{2^{m-1}-1} \sum_{j=-2^{m-1}}^{2^{m-1}-1} \left((v_{j+1,k}^h)^2 + (v_{j,k}^h)^2 + v_{j+1,k}^h v_{j,k}^h\right) + \\ \frac{h^2}{6} \sum_{j=-2^{m-1}}^{2^{m-1}-1} \left((v_{j+1,-2^{m-1}}^h)^2 + (v_{j,-2^{m-1}}^h)^2 + v_{j+1,-2^{m-1}}^h v_{j,-2^{m-1}}^h\right) + \\ \frac{h^2}{6} \sum_{j=-2^{m-1}}^{2^{m-1}-1} \left((v_{j+1,2^{m-1}}^h)^2 + (v_{j,2^{m-1}}^h)^2 + v_{j+1,2^{m-1}}^h v_{j,2^{m-1}}^h\right), \end{cases} \tag{5.14}$$

and we note that the first sum in the right hand side of (5.14) is bounded by the norm of $p_h^0 r_h\, v$ and the second and the last terms vanish as $h \to 0$. In fact,

$$\frac{h^2}{3} \sum_{k=-2^{m-1}+1}^{2^{m-1}-1} \sum_{j=-2^{m-1}}^{2^{m-1}-1} \left((v_{j+1,k}^h)^2 + (v_{j,k}^h)^2 + v_{j+1,k}^h v_{j,k}^h\right) \leq$$

$$\frac{h^2}{3} \sum_{k=-2^{m-1}+1}^{2^{m-1}-1} \sum_{j=-2^{m-1}}^{2^{m-1}-1} \left((v_{j+1,k}^h)^2 + (v_{j,k}^h)^2 + \frac{1}{2}((v_{j+1,k}^h)^2 + (v_{j,k}^h)^2)\right) \leq$$

$$\frac{h^2}{3} \sum_{k=-2^{m-1}+1}^{2^{m-1}-1} \sum_{j=-2^{m-1}}^{2^{m-1}-1} \left((1+1/2)(v_{j+1,k}^h)^2 + (1+1/2)(v_{j,k}^h)^2\right) \leq ||p_h^0 r_h\, v||_{L^2(\Omega)}^2$$

(see (4.6) with $w_h = p_h^0 r_h\, v$) and

$$\frac{h^2}{6} \sum_{j=-2^{m-1}}^{2^{m-1}-1} \left((v_{j+1,2^{m-1}}^h)^2 + (v_{j,2^{m-1}}^h)^2 + v_{j+1,2^{m-1}}^h v_{j,2^{m-1}}^h\right) \leq$$

$$\frac{h^2}{6} \sum_{j=-2^{m-1}}^{2^{m-1}-1} \left((1+1/2)(v_{j+1,2^{m-1}}^h)^2 + (1+1/2)(v_{j,2^{m-1}}^h)^2\right) \leq$$

$$\frac{3h^2}{6} \sum_{j=-2^{m-1}}^{2^{m-1}} (v_{j,2^{m-1}}^h)^2 = \frac{3h^2}{6h^4} \sum_{j=-2^{m-1}}^{2^{m-1}} \left(\int_\Omega v\theta_{j,2^{m-1}} d\,x\right)^2 \leq$$

$$\frac{3h^4}{6h^4} \sum_{j=-2^{m-1}}^{2^{m-1}} \int_\Omega v^2 \theta_{j,2^{m-1}} d\,x = \frac{1}{2} \int_{\Omega_h} v^2 d\,x,$$

where $\Omega_h = (-L, L) \times (L - \frac{h}{2}, L)$. As for any $v \in L^2(\Omega)$, the last integral tends to 0 as $h \to 0$, then from the previous calculations we deduce

$$\limsup ||v_h^1||_{L^2(\Omega)} \leq \limsup ||p_h^0 r_h\, v||_{L^2(\Omega)} = ||v||_{L^2(\Omega)} \tag{5.15}$$

as we have previously proved that the sequence $p_h^0 r_h\, v$ strongly converges to $v$ in $L^2(\Omega)$. This completes the proof of the strong convergence of $v_h^1$ to $v$. As in the same way, we can prove the strong convergence of $v_h^2$ to $v$, the proof of (5.1) is also completed, and this also concludes the proof of condition (a) in Lemma 2. We will prove condition (aa) in Lemma 2 by a density argument. More precisely for a given function $u \in H_0^1(\Omega)$ and any positive $\varepsilon$, let $v \in C_0^1(\Omega)$ such that

$$\max\{||u - v||_{L^2(\Omega)}, ||\frac{\partial u}{\partial x_1} - \frac{\partial v}{\partial x_1}||_{L^2(\Omega)}, ||\frac{\partial u}{\partial x_2} - \frac{\partial v}{\partial x_2}||_{L^2(\Omega)}\} \leq \frac{\varepsilon}{2},$$

then we choose

$$v_h = (v_h^1, v_h^2) = p_h^1\, r_h\, v. \tag{5.16}$$

Of course (for $h$ sufficiently small), the sequence $v_h$ belongs to the space $\in X_{h,0}$, and it converges (in the $X-$ norm) toward the function $u$ in fact

$$||u - v_h^1||^2_{L^2(\Omega)} + ||u - v_h^2||^2_{L^2(\Omega)} + ||\frac{\partial u}{\partial x_1} - \frac{\partial v_h^1}{\partial x_1}||^2_{L^2(\Omega)} + ||\frac{\partial u}{\partial x_2} - \frac{\partial v_h^2}{\partial x_2}||^2_{L^2(\Omega)} \leq$$

$$\varepsilon^2 + ||v - v_h^1||^2_{L^2(\Omega)} + ||v - v_h^2||^2_{L^2(\Omega)} + ||\frac{\partial v}{\partial x_1} - \frac{\partial v_h^1}{\partial x_1}||^2_{L^2(\Omega)} + ||\frac{\partial v}{\partial x_2} - \frac{\partial v_h^2}{\partial x_2}||^2_{L^2(\Omega)},$$

and we use the properties in (5.1) and (5.2) to conclude the proof of condition (aa) and also of Lemma 2.

## 6 Comments

Similar approximation results hold for Sobolev spaces on any plane domain (possibly with a fractal boundary) that enjoys the *extension property*. A domain $\Omega$ enjoys the extension property if there exists a linear operator $E$ from $H^1(\Omega)$ to $H^1(\mathbb{R}^2)$ and a constant $C$ such that

$$||Eu||_{H^1(\mathbb{R}^2)} \leq C||u||_{H^1(\Omega)} \;\forall\; u \in H^1(\Omega). \tag{6.1}$$

For the properties of these domains, we refer to the book of Vladimir Maz'ja [9] and to the papers [7, 8] of Emily Evans.

If the domain $\Omega$ enjoys the extension property, then we can extend every function of $H^1(\Omega)$ over all $\mathbb{R}^2$ and then restrict it to a square $\Omega^*$ (of side $2L$) that contains $\Omega$. On the square $\Omega^*$, we construct the grid $\bar{G}_m$, the operators $r_h,\; p_h^0,\; p_h^1$, and the discrete spaces $V_h(\Omega^*)$, and $X_h(\Omega^*)$ as in Sect. 2. In particular,

$$X_h(\Omega^*) = \{(v^1(x),\, v^2(x))\},$$

where

$$v^1(x) = \sum_{j,k=-2^{m-1}}^{2^{m-1}} v_{j,k}\, \varphi^1_{jk}(x), \qquad v^2(x) = \sum_{j,k=-2^{m-1}}^{2^{m-1}} v_{j,k}\, \varphi^2_{jk}(x) \tag{6.2}$$

(see (2.12)). According to the previous argument, we can prove that the spaces $X_h(\Omega^*)$ $M$-converge to the space $H^1(\Omega^*)$ in $X(\Omega^*)$ (see the condition (b) of Lemma 1 and condition (a) of Lemma 2).

For every $h$ defined in (2.2), $I$, $J$ defined in (2.3), and $(j,k)$ such that $(jh, kh) \in \bar{G}_m$, we put

$$\mathcal{R}(h) = \{(j,k) \text{ such that } int(I \times J) \bigcap \Omega \neq \emptyset\} \text{ and } \Omega^* = int\Big( \bigcup_{(j,k)\in\mathcal{R}(h)} I \times J \Big).$$

We define (for every $h$) the space $X_h(\Omega)$ as the space of the vectors $v_h = (v_h^1, v_h^2)$ whose components are the restrictions to the domain $\Omega$ of the elements $(v_h^{1,*}, v_h^{2,*})$, i.e.,

$$v_{h,\Omega}^1 = v_h^{1,*}|_\Omega \text{ and } v_{h,\Omega}^2 = v_h^{2,*}|_\Omega,$$

where

$$v_h^{1,*}(x) = \sum_{(j,k)\in\mathcal{R}(h)} v_{j,k}\, \varphi_{jk}^1(x), \qquad v_h^{2,*}(x) = \sum_{(j,k)\in\mathcal{R}(h)} v_{j,k}\, \varphi_{jk}^2(x)$$

(see (2.12)).

According to Sect. 2, we set

$$X(\Omega) \equiv H^1(\Omega)_1 \times H^1(\Omega)_2$$

of all pairs $u = (u_1, u_2)$, $u_1 \in H^1(\Omega)_1$, $u_2 \in H^1(\Omega)_2$, which is a Hilbert space with the product norm

$$||u||_{X(\Omega)} = \Big(||u_1||^2_{L^2(\Omega)} + ||\frac{\partial u_1}{\partial x_1}||^2_{L^2(\Omega)} + ||u_2||^2_{L^2(\Omega)} + ||\frac{\partial u_2}{\partial x_2}||^2_{L^2(\Omega)}\Big)^{\frac{1}{2}}. \tag{6.3}$$

Then the spaces $X_h(\Omega)$ $M$-converge to space $H^1(\Omega)$ in $X(\Omega)$. In fact, to show condition (a) of Lemma 2, we denote by $u^* = (Eu)_{|\Omega^*}$, where the operator $E$ is the linear operator from $H^1(\Omega)$ to $H^1(\mathbb{R}^2)$ satisfying (6.1), and we set (see (5.1) and (5.2))

$$p_h^1 r_h u^* = (v_h^1(x),\, v_h^2(x)) \text{ where } v_h^i(x) = \sum_{j,k=-2^{m-1}}^{2^{m-1}} v_{j,k}\, \varphi_{jk}^i(x),$$

$$v_{j,k} = (r_h u^*)_{j,k} \text{ and } i = 1, 2.$$

Then we define

$$p_h^1 r_h u = (v_{h,\Omega}^1(x),\ v_{h,\Omega}^2(x)) \text{ where } v_{h,\Omega}^i = v_h^{i,*}|_\Omega$$

$$v_h^{i,*}(x) = \sum_{(j,k)\in\mathcal{R}(h)} v_{j,k}\,\varphi_{jk}^i(x),\quad v_{j,k} = (r_h u^*)_{j,k} \text{ and } i = 1, 2.$$

We have

$$||u - v_{h,\Omega}^1||_{L^2(\Omega)} \le ||u^* - v_h^1||_{L^2(\Omega^*)} \to 0 \text{ as } h \to 0,\ \forall u \in H^1(\Omega),$$

$$||u - v_{h,\Omega}^2||_{L^2(\Omega)} \le ||u^* - v_h^2||_{L^2(\Omega^*)} \to 0 \text{ as } h \to 0,\ \forall u \in H^1(\Omega),$$

and

$$||\frac{\partial u}{\partial x_1} - \frac{\partial v_{h,\Omega}^1}{\partial x_1}||_{L^2(\Omega)} \le ||\frac{\partial u^*}{\partial x_1} - \frac{\partial v_h^1}{\partial x_1}||_{L^2(\Omega^*)} \to 0 \text{ as } h \to 0,\ \forall u \in H^1(\Omega),$$

$$||\frac{\partial u}{\partial x_2} - \frac{\partial v_{h,\Omega}^2}{\partial x_2}||_{L^2(\Omega)} \le ||\frac{\partial u^*}{\partial x_2} - \frac{\partial v_h^2}{\partial x_2}||_{L^2(\Omega^*)} \to 0 \text{ as } h \to 0,\ \forall u \in H^1(\Omega).$$

Now we briefly discuss condition (b) of Lemma 1. Let the sequence $v_h \subset X_h(\Omega)$ weakly converge to $u = (u^1, u^2)$ in $X(\Omega)$ (see (6.3)), and we recall that the components $v_h = (v_{h,\Omega}^1, v_{h,\Omega}^2)$ are the restrictions to the domain $\Omega$ of the elements

$$v_h^{1,*}(x) = \sum_{(j,k)\in\mathcal{R}(h)} v_{j,k}\,\varphi_{jk}^1(x),\qquad v_h^{2,*}(x) = \sum_{(j,k)\in\mathcal{R}(h)} v_{j,k}\,\varphi_{jk}^2(x). \tag{6.4}$$

We denote by

$$w_h = (w_h^1, w_h^2)$$

the extension by zero of the elements $v_h \in X_h(\Omega)$ i.e.,

$$w_h^1(x) = \sum_{j,k=-2^{m-1}}^{2^{m-1}} w_{j,k}\,\varphi_{jk}^1(x),\qquad v_h^2(x) = \sum_{j,k=-2^{m-1}}^{2^{m-1}} w_{j,k}\,\varphi_{jk}^2(x) \tag{6.5}$$

where for $(j, k)$ such that $(jh, kh) \in \bar{G}_m$ we put

$$w_{j,k} = \begin{cases} v_{j,k} & \text{in (6.4)} \quad \text{if } (j,k) \in \mathcal{R}(h) \\ 0 & \text{otherwise.} \end{cases} \tag{6.6}$$

We note that the sequence $w_h$ is bounded in the space $X(\Omega^*)$ and then (up to pass to a subsequence) weakly converges to $w = (w^1, w^2)$ in $X(\Omega^*)$.

In Sect. 2 we proved that $w^1 = w^2$ and $w$ belongs to the space $H^1(\Omega^*)$ (the diagonal of $X(\Omega^*)$). Finally, as in the set $\Omega$ we have $w^1 = u^1$ and $w^2 = u^2$, then $u$ belongs to the space $H^1(\Omega)$.

*Fractal* examples of domains with the extension property are the generalized Koch Islands $\Omega_\alpha$ for any $\alpha \in (2, 4)$. These domains are domains in $\mathbb{R}^2$ having as sides Koch curves $K_\alpha$ that can be constructed starting by a regular polygon and replacing each side by a Koch curve. The Koch curve $K_\alpha$ (with endpoints $A = (0, 0)$, and $B = (1, 0)$) is obtained as the compact invariant set associated with the family of the four contractive similarities with contraction factor $\alpha^{-1}$, $\alpha \in (2, 4)$ : $\Psi_\alpha = \{\psi_{1,\alpha} \ldots, \psi_{4,\alpha}\}$, where $\psi_{i,\alpha} : \mathbb{C} \to \mathbb{C}$, $i = 1, \ldots, 4$,

$$\psi_{1,\alpha}(z) = \frac{z}{\alpha}, \qquad \psi_{2,\alpha}(z) = \frac{z}{\alpha} e^{i\theta(\alpha)} + \frac{1}{\alpha},$$

$$\psi_{3,\alpha}(z) = \frac{z}{\alpha} e^{-i\theta(\alpha)} + \frac{1}{2} + i\sqrt{\frac{1}{\alpha} - \frac{1}{4}}, \qquad \psi_{4,\alpha}(z) = \frac{z-1}{\alpha} + 1,$$

$$\theta(\alpha) = \arcsin\left(\frac{\sqrt{\alpha(4-\alpha)}}{2}\right). \tag{6.7}$$

Other examples of domains that enjoy the extension property are the pre-fractal Koch Islands $\Omega_\alpha^n$ that are the polygonal domains having as sides pre-fractal Koch curves (see Fig. 3 below). The pre-fractal Koch curve $K_\alpha^n$ (with endpoints $A = (0, 0)$, and $B = (1, 0)$) can be constructed (for any fixed $n \in \mathbb{N}$) stopping at the step $n$ the iteration procedure that involves the family of the four contractive similarities: $\Psi_\alpha = \{\psi_{1,\alpha} \ldots, \psi_{4,\alpha}\}$. Let $K^0$ be the line segment of unit length that

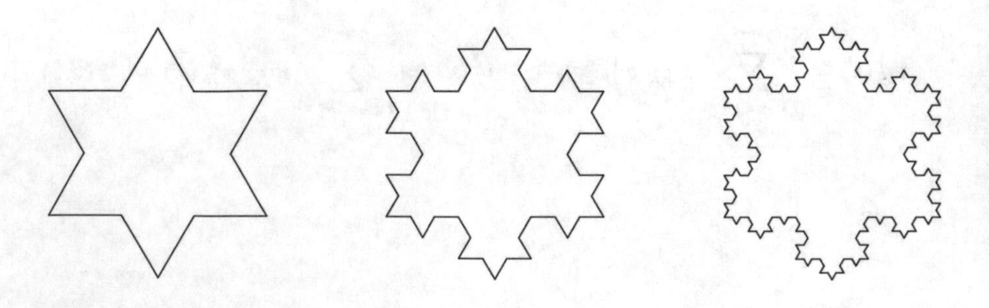

**Fig. 3** $\alpha = 3$

has as endpoints $A = (0,0)$ and $B = (1,0)$. We set

$$K_\alpha^1 = \bigcup_{i=1}^{4} \psi_{i,\alpha}(K^0), \qquad K_\alpha^2 = \bigcup_{i=1}^{4} \psi_{i,\alpha}(K_\alpha^1) = \bigcup_{i|2} \psi_{i|2,\alpha}(K^0), \qquad \ldots,$$

$$K_\alpha^n = \bigcup_{i=1}^{4} \psi_{i,\alpha}(K_\alpha^{n-1}) = \bigcup_{i|n} \psi_{i|n,\alpha}(K^0).$$

Incidentally we note that the domains $\Omega_\alpha^n$ admit extension operators $E_n$ uniformly bounded in $n$. Therefore, they satisfy the estimate (6.1) with a constant $C$ independent of $n$ (see for instance [4] and [5]).

Finally, we point out that our results in this paper are quite novel in the literature on the subject. In fact, all approximation results for Sobolev spaces available so far—including those in [6] and [1] mentioned in the introduction—require the domain to have a very regular boundary, while we only require the domain to have the extension property, what includes domains with a very irregular and possibly fractal boundary, as explained before.

## References


1. Aubin, J.P.: Approximation of elliptic boundary-value problems. In: Courant, R., Bers, L., Stoker, J.J. (eds.) Wiley-Interscience, Hoboken (1972)
2. Barbu, V.: Self-organized criticality and convergence to equilibrium of solutions to nonlinear diffusion problems. Annu. Rev. Control. **340**, 52–61 (2010)
3. Barbu, V.: Self-organized criticality of cellular automata model; absorbtion in finite-time of supercritical region into the critical one. Math. Meth. Appl. Sci. **36**, 1726–1733 (2013)
4. Capitanelli, R.: Asymptotics for mixed Dirichlet-Robin problems in irregular domains. J. Math. Anal. Appl. **362**(2), 450–459 (2010)
5. Capitanelli, R., Vivaldi, M.A.: Reinforcement problems for variational inequalities on fractal sets. Calc. Var. Partial Differ. Equ. **54**(3), 2751–2783 (2015)
6. Céa, J.: Approximation variationnelle des problèmes aux limites. Ann. Inst. Fourier (Grenoble) **14**(2), 345–444 (1964)
7. Evans, E.: A finite element approach to H1 extension using prefractals. Adv. Math. Sci. Appl. **22**(2), 391–420 (2012)
8. Evans, E.: A finite element approach to Hölder extension using prefractals. Meth. Appl. Anal. **19**(2), 161–186 (2012)
9. Maz'ja, V.G.: Sobolev spaces. Translated from the Russian by T. O. Shaposhnikova. Springer Series in Soviet Mathematics. Springer, Berlin (1985)
10. Mosco, U.: Convergence of convex sets and of solutions of variational inequalities. Adv. Math. **3**, 510–585 (1969)
11. Mosco, U.: An Introduction to the approximate solution of variational inequalities. In: Geymonat, G. (ed.) Constructive Aspects of Functional Analysis, pp. 497–685. Lectures given at a Summer School of the Centro Internazionale Matematico Estivo (C.I.M.E.) Erice (Trapani), June 27-July 7 (1971)

12. Mosco, U.: Between discrete and continuous structures. Lectures given at Riemann International School of Mathematics. Villa Toeplitz - Varese, June 19–23 (2017)
13. Mosco, U.: Finite-time self-organized-criticality on synchronized infinite grids. SIAM J. Math. Anal. **50**, 2409–2440 (2018)
14. Mosco, U., Vivaldi, M.A.: On a discrete self-organized-criticality finite time result. Discrete Contin. Dynam. Systems **40**(8), 5079–5103 (2020). https://doi.org/10.3934/dcds.2020212

# Generalization of Rellich–Kondrachov Theorem and Trace Compactness for Fractal Boundaries

**Anna Rozanova-Pierrat**

**Abstract** We present a survey of recent results from the functional analysis that allow to solve PDEs in a large class of domains with irregular boundaries. We extend the previously introduced concept of admissible domains with a $d$-set boundary to domains with boundaries carrying measures that are not necessarily Ahlfors $d$-regular. We prove the generalizations of the Rellich–Kondrachov theorem and the compactness of the trace operator and obtain uniqueness and existence results for weak solutions to Poisson boundary value problems with Robin boundary conditions. We observe the usual properties of the associated spectral problem.

**Keywords** Fractal boundaries · Compact operators · $d$-set · Trace and extension operators · Rellich–Kondrachov theorem

## 1 Introduction

From the theory of partial differential equations, it is known that the irregularity of the boundary of the considered domain can be a serious obstacle even for the proof of the existence of a weak solution. In this paper we are interested to identify a class of domains with irregular boundaries for which we still have weak well-posedness for elliptic problems. Here we consider a specific problem, namely the Poisson equation with homogeneous Robin boundary conditions,

$$\begin{cases} -\Delta u = f \text{ in } \Omega, \\ \frac{\partial u}{\partial \nu} + \alpha u = 0 \text{ with } \alpha > 0 \text{ on } \partial\Omega. \end{cases} \tag{1}$$

To provide a rigorous weak formulation of this problem, it is important to be able to integrate by parts and to have a well-defined trace operator for $\partial\Omega$. For

A. Rozanova-Pierrat (✉)
CentraleSupélec, Université Paris-Saclay, Gif-sur-Yvette, France
e-mail: anna.rozanova-pierrat@centralesupelec.fr



M. R. Lancia, A. Rozanova-Pierrat (eds.), *Fractals in Engineering: Theoretical Aspects and Numerical Approximations*, SEMA SIMAI Springer Series 8,
https://doi.org/10.1007/978-3-030-61803-2_7

smooth or Lipschitz boundaries $\partial\Omega$, this is classical and well-known (for sufficiently smooth boundary, see Raviart and Thomas [35], and for the Lipschitz case, see Marschall [32] and [15, 34]). If $\partial\Omega$ is Lipschitz, then the normal unit vector $\nu$ to the boundary $\partial\Omega$ exists almost everywhere, and the trace operator $\text{Tr} : W^{1,2}(\Omega) \to H^{\frac{1}{2}}(\partial\Omega)$ is linear, continuous, and surjective, [15, 30, 32, 34], and has a linear right continuous inverse, i.e. an extension operator $E : H^{\frac{1}{2}}(\partial\Omega) \to W^{1,2}(\Omega)$ is such that $\text{Tr}(E(u)) = u$.

Moreover, for $u, v \in W^{1,2}(\Omega)$ with $\Delta u \in L^2(\Omega)$, the usual Green formula holds in the sense that

$$\int_\Omega \nabla u v \mathrm{d}x = \langle \frac{\partial u}{\partial \nu}, \text{Tr} v\rangle_{((H^{\frac{1}{2}}(\partial\Omega))', H^{\frac{1}{2}}(\partial\Omega))} - \int_\Omega \nabla v \nabla u \mathrm{d}x. \quad (2)$$

This formula understands the existence of the normal derivative of $u$ on $\partial\Omega$ as the existence of a linear continuous form on $H^{\frac{1}{2}}(\partial\Omega)$, where $H^{\frac{1}{2}}(\partial\Omega)$ is the image of $W^{1,2}(\Omega)$ for a Lipschitz domain $\Omega$ by the trace operator. The dual space $(H^{\frac{1}{2}}(\partial\Omega))'$ is usually denoted by $H^{-\frac{1}{2}}(\partial\Omega)$.

For Lipschitz domains, it is also possible to give a definition in the weak sense of the divergence operator for vector valued functions (see, for instance, Theorem 2.5 § 2 [14]) and to introduce the usual formula of integration by parts for all $u$ and $v$ from $W^{1,2}(\Omega)$ in the following weak sense:

$$\langle u\nu_i, v\rangle_{(H^{-\frac{1}{2}}(\partial\Omega), H^{\frac{1}{2}}(\partial\Omega))} := \int_\Omega \frac{\partial u}{\partial x_i} v \mathrm{d}x + \int_\Omega u \frac{\partial v}{\partial x_i} \mathrm{d}x \quad i = 1, \ldots, n, \quad (3)$$

where $u\nu_i$ is denoted as the linear continuous functional on $H^{\frac{1}{2}}(\partial\Omega)$.

Thanks to the classical results of Calderon [8] and Stein [37], it is known that every Lipschitz domain $\Omega$ is an extension domain for the Sobolev space $W^{k,p}(\Omega)$ with $1 \le p \le \infty$, $k \in \mathbb{N}^*$, and we briefly recall the definition of this notion.

**Definition 1 ($W^{k,p}$-Extension Domains)** A domain $\Omega \subset \mathbb{R}^n$ is called a $W^{k,p}$-extension domain ($k \in \mathbb{N}^*$) if there exists a bounded linear extension operator $E : W^{k,p}(\Omega) \to W^k_p(\mathbb{R}^n)$,

*i.e.* a linear operator assigning a function $v = Eu \in W^k_p(\mathbb{R}^n)$ with $v|_\Omega = u$ to any $u \in W^{k,p}(\Omega)$, such that

$$\|v\|_{W^k_p(\mathbb{R}^n)} \le C\|u\|_{W^{k,p}(\Omega)} \quad \text{with a universal constant } C > 0.$$

This result was generalized by Jones [20] in the framework of $(\varepsilon, \infty)$-domains that give an optimal class of extension domains in $\mathbb{R}^2$, but not in $\mathbb{R}^3$. The optimal class of extension domains for $p > 1$ in $\mathbb{R}^n$ was found more recently by Hajłas, Koskela and Tuominen [16]. These results are discussed in Sect. 2, where we give all definitions.

Thanks to the results in [4, 5, 25, 27, 40], it is possible to generalize the trace operator to cases of more irregular boundaries, such as $d$-sets, and even to sets without a fixed dimension [18, 22]. The definition of the trace for a regular distribution and different image spaces leading to different Green formulas is presented in Sect. 3.

But to be able to ensure the weak well-posedness of problem (1) and to discuss the associated spectral problem for $-\Delta$, we also need the compactness of the inclusion $W^{1,2}(\Omega)$ in $L^2(\Omega)$ and the compactness of the trace operator, now considered as an operator from $W^{1,2}(\Omega)$ to $L^2(\partial\Omega)$.

Thanks to [12] Theorem V.4.17, it is known that if a domain $\Omega$ has a continuous boundary (in the sense of graphs, see [12] Definition V.4.1), then $W^{1,2}(\Omega)$ is compactly embedded in $L^2(\Omega)$. General $d$-set boundaries with $d > n - 1$, as for instance a von Koch curve, do not satisfy the assumption to have a continuous boundary. In our article [4] we proved this fact in the framework of admissible domains with a $d$-set boundary. Here we also prove it for more general boundaries as in [21, 22] (see Sect. 4). This may be seen as an update to the concept of admissible domains introduced first in [4]. We follow the same idea as in [4] and introduce the class of all Sobolev extension domains with boundaries on which one can define a surjective linear continuous trace operator with linear continuous right inverse. To emphasize their "extension nature," we call these domains Sobolev admissible domains (see Definition 7).

The most common examples of Sobolev admissible domains are domains with regular or Lipschitz boundaries, with $d$-set boundaries, such as Von Koch fractals or with a "mixed" boundary (for instance, the boundary of a cylindrical domain in $\mathbb{R}^3$ with the base being a snowflake domain in $\mathbb{R}^2$ as considered in [10, 29]).

The generalization of the Kondrachov–Rellich theorem in the framework of Sobolev admissible domains allows to extend the compactness studies of the trace from [3] and to update the results of [4] (see Sect. 5): for a Sobolev admissible domain with a compact boundary, the trace operator mapping from $W^{1,2}(\Omega)$ to $L^2(\partial\Omega)$ is compact.

Thus, as for the usual Lipschitz bounded case, the problem (1) is weakly well-posed and the corresponding spectral problem has a countable number of eigenvalues going to $+\infty$ with the eigenfunctions forming an orthogonal basis in $W^{1,2}(\Omega)$ that becomes an orthonormal basis in $L^2(\Omega)$ by the classical Hilbert–Schmidt theorem for compact self-adjoint operators on a Hilbert space (see Sect. 6).

The paper is organized as follows. In Sect. 2 we present recent results on Sobolev extension domains. In Sect. 3 we firstly define the trace operator on a $d$-set in Sect. 3.1, and secondly, in Sect. 3.2 we provide analogous results in a framework that does not require the boundaries to be exactly $d$-dimensional. We finish the section by a generalization of the Green formula and the integration by parts formula within this framework in Sect. 3.3. Using the results on the trace and on the extension operators, we introduce the concept of Sobolev admissible domains in Sect. 4 and generalize the Rellich–Kondrachov theorem. In Sect. 5 we show the compactness of the trace operator considered as an operator mapping to $L^p(\partial\Omega)$. In Sect. 6 we apply these theorems to show the well-posedness of the Poisson problem (1) on the $W^{1,2}$-Sobolev admissible domains.

## 2 Sobolev Extension Domains

As in [4], we start by recalling the classical results of Calderon [8] and Stein [37]: every Lipschitz domain $\Omega$ is an extension domain for $W^{k,p}(\Omega)$ with $1 \leq p \leq \infty$, $k \in \mathbb{N}^*$. This result was generalized by Jones [20] in the framework of $(\varepsilon, \delta)$-domains.

**Definition 2 ($(\varepsilon, \delta)$-Domain [20, 23, 40])** An open connected subset $\Omega$ of $\mathbb{R}^n$ is an $(\varepsilon, \delta)$-domain, $\varepsilon > 0$, $0 < \delta \leq \infty$, if whenever $x, y \in \Omega$ and $|x - y| < \delta$, there is a rectifiable arc $\gamma \subset \Omega$ with length $\ell(\gamma)$ joining $x$ to $y$ and satisfying

1. $\ell(\gamma) \leq \frac{|x-y|}{\varepsilon}$ (thus locally quasiconvex) and
2. $d(z, \partial\Omega) \geq \varepsilon|x - z|\frac{|y-z|}{|x-y|}$ for $z \in \gamma$.

The constant $\delta$ is allowed to equal $+\infty$. For $\delta = +\infty$, it is possible to avoid the local character of this definition, and in this case $\Omega$ is said to be an $(\varepsilon, \infty)$-domain. Definition 2 without the second condition yields the definition of a locally quasiconvex domain. The second condition prohibits the boundary to collapse into thin structures. This is in stark contrast to the case of fractal trees, in fact, this is the reason why fractal trees [1] are not $(\varepsilon, \infty)$-domains.

The $(\varepsilon, \delta)$-domains are also called locally uniform domains [17]. Actually, bounded locally uniform domains, or bounded $(\varepsilon, \delta)$-domains, are equivalent (see [17] point 3.4) to the uniform domains, firstly defined by Martio and Sarvas in [33], for which there are no more restriction $|x - y| < \delta$ (see Definition 2).

Thanks to Jones [20], it is known that any $(\varepsilon, \delta)$-domain in $\mathbb{R}^n$ is a $W^{k,p}$-extension domain for all $1 \leq p \leq \infty$ and $k \in \mathbb{N}^*$. Moreover, for a bounded finitely connected domain $\Omega \subset \mathbb{R}^2$, Jones [20] proved that

$$\begin{aligned}\Omega \text{ is a } W^{k,p}\text{-extension domain } (1 \leq p \leq \infty \text{ and } k \in \mathbb{N}^*) &\Longleftrightarrow \\ \Omega \text{ is an } (\varepsilon, \infty)\text{-domain for some } \varepsilon > 0 &\Longleftrightarrow \\ \text{the boundary } \partial\Omega \text{ consists of finite number of points and quasi-circles.}\end{aligned}$$

However, it is no more true for $n \geq 3$, *i.e.* there are $W^{1,p}$-extension domains that are not locally uniform [20] (in addition, an $(\varepsilon, \delta)$-domain in $\mathbb{R}^n$ with $n \geq 3$ is not necessarily a quasi-sphere).

To discuss general properties of locally uniform domains, let us introduce Ahlfors $d$-regular sets, which are more currently called by $d$-sets.

**Definition 3 (Ahlfors $d$-Regular Set or $d$-Set [23, 24, 38, 40])** Let $F$ be a closed Borel non-empty subset of $\mathbb{R}^n$. The set $F$ is called a $d$-set ($0 < d \leq n$) if there exists a $d$-measure $\mu$ on $F$, i.e. a positive Borel measure with support $F$ ($\operatorname{supp} \mu = F$) such that there exist constants $c_1, c_2 > 0$,

$$c_1 r^d \leq \mu(\overline{B_r(x)}) \leq c_2 r^d, \quad \text{for } \forall\, x \in F,\ 0 < r \leq 1,$$

where $B_r(x) \subset \mathbb{R}^n$ denotes the Euclidean ball centered at $x$ and of radius $r$.

As [23, Prop. 1, p 30] all $d$-measures on a fixed $d$-set $F$ are equivalent, it is also possible to define a $d$-set by the $d$-dimensional Hausdorff measure $m_d$

$$c_1 r^d \le m_d(F \cap \overline{B_r(x)}) \le c_2 r^d, \quad \text{for } \forall\, x \in F,\ 0 < r \le 1,$$

which in particular implies that $F$ has Hausdorff dimension $d$ in the neighborhood of each point of $F$ [23, p.33].

If the boundary $\partial\Omega$ is a $d$-set endowed with the $d$-dimensional Hausdorff measure restricted to $\partial\Omega$, then we denote by $L^p(\partial\Omega, m_d)$ the Lebesgue space defined with respect to this measure with the norm

$$\|u\|_{L^p(\partial\Omega,m_d)} = \left(\int_{\partial\Omega} |u|^p dm_d\right)^{\frac{1}{p}}.$$

From [40], it is known that

- All $(\varepsilon, \delta)$-domains in $\mathbb{R}^n$ are $n$-sets ($d$-set with $d = n$)

  $$\exists c > 0 \quad \forall x \in \overline{\Omega},\ \forall r \in ]0, \delta[\cap]0, 1] \quad \lambda(B_r(x) \cap \Omega) \ge C\lambda(B_r(x)) = cr^n,$$

  where $\lambda(A)$ denotes the Lebesgue measure of a set $A$ in $\mathbb{R}^n$. This property is also called the measure density condition [16]. Let us notice that an $n$-set $\Omega$ cannot be "thin" close to its boundary $\partial\Omega$, since it must all times contain a non trivial ball in its neighborhood.
- If $\Omega$ is an $(\varepsilon, \delta)$-domain and $\partial\Omega$ is a $d$-set ($d < n$), then $\overline{\Omega} = \Omega \cup \partial\Omega$ is an $n$-set.

In particular, a Lipschitz domain $\Omega$ of $\mathbb{R}^n$ is an $(\varepsilon, \delta)$-domain and also an $n$-set [40]. But not every $n$-set is an $(\varepsilon, \delta)$-domain: adding an in-going cusp to an $(\varepsilon, \delta)$-domain, we obtain an $n$-set that is not an $(\varepsilon, \delta)$-domain anymore. Classical snowflake domains are examples of $(\varepsilon, \infty)$-domains with $d$-set boundary [9, 40], $d > n - 1$.

Recently, Hajłasz, Koskela, and Tuominen [16] have proved that every $W^{k,p}$-extension domain in $\mathbb{R}^n$ for $1 \le p < \infty$ and $k \ge 1$, $k \in \mathbb{N}$ is an $n$-set. In addition they proved the following statements.

**Theorem 1**

*(i) A domain $\Omega \subset \mathbb{R}^n$ is a $W^1_\infty$-extension domain if and only if $\Omega$ is uniformly locally quasiconvex.*

*(ii) For $1 < p < \infty$, $k = 1, 2, \ldots$, a domain $\Omega \subset \mathbb{R}^n$ is a $W^k_p$-extension domain if and only if $\Omega$ is an $n$-set and $W^{k,p}(\Omega) = C^k_p(\Omega)$ (in the sense of equivalent norms).*

By $C^k_p(\Omega)$ is denoted the space of the fractional sharp maximal functions.

**Definition 4** For a set $\Omega \subset \mathbb{R}^n$ of positive Lebesgue measures,

$$C_p^k(\Omega) = \{f \in L^p(\Omega)|$$

$$f_{k,\Omega}^\sharp(x) = \sup_{r>0} r^{-k} \inf_{P\in\mathcal{P}^{k-1}} \frac{1}{\lambda(B_r(x))} \int_{B_r(x)\cap\Omega} |f - P| \mathrm{d}y \in L^p(\Omega)\}$$

with the norm $\|f\|_{C_p^k(\Omega)} = \|f\|_{L^p(\Omega)} + \|f_{k,\Omega}^\sharp\|_{L^p(\Omega)}$. By $\mathcal{P}^{k-1}$ we denote the space of polynomials of the order $k-1$.

From [20] and [16], we immediately obtain the following [4].

**Corollary 1** *Let $\Omega$ be a bounded finitely connected domain in $\mathbb{R}^2$ and $1 < p < \infty$, $k \in \mathbb{N}^*$. The domain $\Omega$ is a 2-set with $W^{k,p}(\Omega) = C_p^k(\Omega)$ (with norms' equivalence) if and only if $\Omega$ is an $(\varepsilon, \delta)$-domain and its boundary $\partial\Omega$ consists of a finite number of points and quasi-circles.*

The question about $W_p^k$-extension domains is equivalent to the question of the continuity of the trace operator $\mathrm{Tr} : W_p^k(\mathbb{R}^n) \to W_p^k(\Omega)$, the trace operator on the domain $\Omega$. In the next section we introduce the notion of trace to more general Borel sets, and we will use this notion of trace to study boundary conditions.

# 3 Trace on the Boundary and Green Formulas

## 3.1 *Framework of d-Sets and Markov's Local Inequality*

From [23] p.39, it is also known that all closed $d$-sets with $d > n - 1$ preserve Markov's local inequality.

**Definition 5 (Markov's Local Inequality)** A closed subset $V$ in $\mathbb{R}^n$ preserves Markov's local inequality if for every fixed $j \in \mathbb{N}^*$, there exists a constant $c = c(V, n, j) > 0$, such that

$$\max_{V\cap\overline{B_r(x)}} |\nabla P| \leq \frac{c}{r} \max_{V\cap\overline{B_r(x)}} |P|$$

for all polynomials $P \in \mathcal{P}_j$ and all closed balls $\overline{B_r(x)}$, $x \in V$ and $0 < r \leq 1$.

For instance, self-similar sets that are not subsets of any $(n-1)$-dimensional subspace of $\mathbb{R}^n$, the closure of a domain $\Omega$ with Lipschitz boundary, and also $\mathbb{R}^n$ itself preserve Markov's local inequality (see Refs. [25, 40]). The geometrical characterization of sets preserving Markov's local inequality was initially given in [26] (see Theorem 1.3) and can be simply interpreted as sets that are not too flat anywhere. It can be illustrated by the following theorem of Wingren [41].

**Theorem 2** *A closed subset $V$ in $\mathbb{R}^n$ preserves Markov's local inequality if and only if there exists a constant $c > 0$ such that for every ball $B_r(x)$ centered in $x \in V$ and with the radius $0 < r \leq 1$, there are $n+1$ affinely independent points $y_i \in V \cap B_r(x)$, $i = 1, \dots, n+1$, such that the n-dimensional ball inscribed in the convex hull of $y_1, y_2, \dots, y_{n+1}$, has radius not less than $cr$.*

Smooth manifolds in $\mathbb{R}^n$ of dimension less than $n$, as for instance a sphere, are examples of "flat" sets not preserving Markov's local inequality. More precisely, the sets $F$ that do not preserve Markov's inequality [23, Thm. 2, p.38] are exactly the sets satisfying the geometric condition in the following theorem.

**Theorem 3** *A closed, non-empty subset $F$ of $\mathbb{R}^n$ preserves Markov's inequality if and only if the following geometric condition does not hold: for every $\varepsilon > 0$, there exists a ball $B_r(x_0)$, $x_0 \in F$, $0 < r \leq 1$, so that $B_r(x_0) \cap F$ is contained in some band of type $\{x \in \mathbb{R}^n |\ (b, x - x_0)_{\mathbb{R}^n} < \varepsilon r\}$, where $b \in \mathbb{R}^n$, $|b| = 1$, and $(b, x - x_0)_{\mathbb{R}^n}$ is the scalar product of $b$ and $x - x_0$.*

Relationships between Markov inequalities and inequalities of Sobolev–Gagliardo–Nirenberg type were studied in [7]. The advantage of $d$-set boundaries preserving Markov's inequality (thus $0 < d < n$) is that, [39, 2.1], there exists a bounded linear extension operator $\hat{E}$ of the Hölder space $C^{k-1,\alpha-k+1}(\partial\Omega)$ to the Hölder space $C^{k-1,\alpha-k+1}(\mathbb{R}^n)$, where for $k \in \mathbb{N}^*$ $k-1 < \alpha \leq k$ (see also [23, p. 2]). This allows to show the existence of a linear continuous extension from the Besov space $B^{p,p}_\alpha(\partial\Omega)$ on $\partial\Omega$ to the Sobolev space $W^k_p(\mathbb{R}^n)$ with $\alpha = k - \frac{(n-d)}{p} \geq 1$ and $k \geq 2$ [25]. For the extensions of minimal regularity with $k = 1$, and thus with $\alpha < 1$ (see in addition the definition of the Besov space Def. 3.2 in [19] with the help of the normalized local best approximation in the class of polynomials $P_{k-1}$ of the degree equal to $k-1$), Markov's inequality is trivially satisfied for $j = 0$ in Definition 5 on all closed sets of $\mathbb{R}^n$, and hence we do not need to impose it [25, p. 198].

Before coming to details of the mentioned results, let us generalize the notion of the trace.

**Definition 6** For an arbitrary open set $\Omega$ of $\mathbb{R}^n$, the trace operator Tr is defined [23] for $u \in L^1_{loc}(\Omega)$ by

$$\mathrm{Tr}u(x) = \lim_{r \to 0} \frac{1}{\lambda(\Omega \cap B_r(x))} \int_{\Omega \cap B_r(x)} u(y)dy,$$

where $\lambda$ denotes the Lebesgue measure on $\mathbb{R}^n$. The trace operator Tr is considered for all $x \in \overline{\Omega}$ for which the limit exists.

Using this definition of the trace, one can prove a trace theorem on closed $d$-sets [23] Ch.VII and [40] Proposition 4, in which we think it should be made more precise that the closed set $F$ should preserve Markov's local inequality not necessarily for all $k \in \mathbb{N}^*$, but at least up to $k-1$ with $k \in \mathbb{N}^*$, the fixed regularity of the Sobolev space of which we take the trace on $F$.

**Theorem 4** *Let $F$ be a closed $d$-set preserving Markov's local inequality at least up to $k-1$ for a fixed $k \in \mathbb{N}^*$. If*

$$0 < d < n, \quad 1 < p < \infty, \quad \text{and} \quad \alpha = k - \frac{(n-d)}{p} > 0,$$

*then the trace operator* $\mathrm{Tr} : W^k_p(\mathbb{R}^n) \to B^{p,p}_\alpha(F)$ *is bounded linear surjection with a bounded right inverse* $E : B^{p,p}_\alpha(F) \to W^k_p(\mathbb{R}^n)$, *i.e.* $\mathrm{Tr} \circ E = \mathrm{Id}$ *on* $B^{p,p}_\alpha(F)$.

The definition of the Besov space $B^{p,p}_\alpha(F)$ on a closed $d$-set $F$ can be found, for instance, in Ref. [23] p. 135 and Ref. [40]. See also Triebel for equivalent definitions [38].

Thanks to [26, Thm. 1.2, p. 145], we notice that, if the Markov inequality is preserved by a set $F$ for $j = 1$, then it is preserved by $F$ for all $j \in \mathbb{N}^*, j \geq 1$. Thus the added condition in Theorem 4 for these sets, especially for $d$-sets with $d > n-1$, is useless. Nevertheless, for the case of $W^1_p$ with $1 < p < \infty$, the theorem explicitly holds for all $n-1$-sets, also not satisfying the Markov inequality with $j = 1$. Note that for $d = n-1$, as it is also mentioned in [5], one has $\alpha = \frac{1}{2}$ and $B^{2,2}_{\frac{1}{2}}(F) = H^{\frac{1}{2}}(F)$ as usual in the case of the classical results [30, 32] for Lipschitz boundaries $\partial\Omega = F$. Since $\alpha = \frac{1}{2} < 1$, as noticed previously the geometrical condition for the boundary to preserve Markov's inequality does not occur.

Moreover, considering only $W^{1,2}(\mathbb{R}^n) = \{u \in L^2(\mathbb{R}^n) |\ \nabla u \in L^2(\mathbb{R}^n)\}$, we deduce the following from Theorem 4.

**Theorem 5** *Let $F$ be a closed $d$-set,*

$$0 \leq n-2 < d < n, \quad \text{and} \quad \alpha = 1 - \frac{(n-d)}{2} > 0.$$

*Then the trace operator* $\mathrm{Tr} : W^{1,2}(\mathbb{R}^n) \to B^{2,2}_\alpha(F)$ *is bounded linear surjection with a bounded right inverse* $E : B^{2,2}_\alpha(F) \to W^{1,2}(\mathbb{R}^n)$, *i.e.* $\mathrm{Tr} \circ E = \mathrm{Id}$ *on* $B^{2,2}_\alpha(F)$.

## 3.2 General Framework of Closed Subsets of $\mathbb{R}^n$

It is possible to consider more general measures than $d$-dimensional measures that can describe by their supports a boundary of a domain [10, 21, 22].

We follow [21, Section 1] and say that a Borel measure $\mu$ on $\mathbb{R}^n$ with support $\operatorname{supp} \mu = F$ *satisfies the $D_s$-condition* for an exponent $0 < s \leq n$ if there is a constant $c_s > 0$ such that

$$\mu(B_{kr}(x)) \leq c_s k^s \mu(B_r(x)), \quad x \in F, \quad r > 0, \quad k \geq 1, \quad 0 < kr \leq 1. \tag{4}$$

Here as previously $B_r(x) \subset \mathbb{R}^n$ denotes an open ball centered at $x$ and of radius $r$. We say that $\mu$ *satisfies the* $L_d$*-condition* for an exponent $0 \leq d \leq n$ if for some constant $c > 0$, we have

$$\mu(B_{kr}(x)) \geq c_d k^d \mu(B_r(x)), \quad x \in F, \quad r > 0, \quad k \geq 1, \quad 0 < kr \leq 1. \tag{5}$$

We also introduce the so-called normalization condition

$$c_1 \leq \mu(B_1(x)) \leq c_2, \quad x \in F, \tag{6}$$

where $c_1 > 0$ and $c_2 > 0$ are the constants independent of $x$.

Combining (4) and (6), one can find a constant $c > 0$ such that

$$\mu(B_r(x)) \geq c\, r^s, \quad x \in F, \quad 0 < r \leq 1, \tag{7}$$

what implies $\dim_H F \leq s$, where $\dim_H F$ denotes the Hausdorff dimension of $F$. Similarly (5) and (6) yield a constant $c' > 0$ such that

$$\mu(B_r(x)) \leq c'\, r^d, \quad x \in F, \quad 0 < r \leq 1, \tag{8}$$

hence $\dim_H F \geq d$. Moreover, (4) implies the doubling condition

$$\mu(B_{2r}(x)) \leq c\, \mu(B_r(x)), \quad x \in F, \quad 0 < r \leq 1/2,$$

where $c > 0$ is a suitable constant [21, Section 1].

If a Borel measure $\mu$ with support $F$ satisfies (7) and (8) with $s = d$ for some $0 < d \leq n$, then, according to Definition 3, $\mu$ is called a *d-measure* and $F$ is called a *d-set*. Obviously, if we have (4), (5), and (6) and $d = s$, then $\mu$ is a $d$-measure and $F$ a $d$-set. Otherwise, we consider measures, which by (7) and (8) satisfy for some constants $c > 0$ and $c' > 0$

$$c\, r^s \leq \mu(B_r(x)) \leq c'\, r^d, \quad x \in F, \quad 0 < r \leq 1. \tag{9}$$

For this general measure $\mu$ supported on a closed subset $F \subset \mathbb{R}^n$, it is possible thanks to [21] to define the corresponding Lebesgue spaces $L^p(F, \mu)$ and Besov spaces $B^{p,p}_\beta(F, \mu)$ on closed subsets $F \subset \mathbb{R}^n$ in such a way that we have the following theorem.

**Theorem 6** *Let* $0 \leq d \leq n$, $d \leq s \leq n$, $s > 0$, $1 \leq p \leq +\infty$,

$$\frac{n-d}{p} < \beta < 1 + \frac{n-s}{p}, \tag{10}$$

*and let* $F \subset \mathbb{R}^n$ *be a closed set that is the support of a Borel measure* $\mu$ *satisfying (4), (5), and (6).*

*Then, considering the Besov space $B^{p,p}_\beta(F,\mu)$ on $F$, defined as the space of $\mu$-classes of real-valued functions $f$ on $F$ such that the norm*

$$\|f\|_{B^{p,p}_\beta(F,\mu)} :=$$
$$\|f\|_{L^p(F,\mu)}+\left(\sum_{\nu=0}^{\infty}2^{\nu(\beta-\frac{n}{p})}\int\int_{|x-y|<2^{-\nu}}\frac{|f(x)-f(y)|^p}{\mu(B(x,2^{-\nu}))\mu(B(y,2^{-\nu}))}\mu(dy)\mu(dx)\right)^{1/p}$$

*is finite, the following statements hold.*

*(i)* $\mathrm{Tr}_F$ *is a continuous linear operator from* $W^\beta_p(\mathbb{R}^n)$ *onto* $B^{p,p}_\beta(F)$*, and*

$$\|\mathrm{Tr}_F\, f\|_{B^{p,p}_\beta(F)} \le c_\beta\, \|f\|_{W^\beta_p(\mathbb{R}^n)}, \quad f\in W^\beta_p(\mathbb{R}^n), \tag{11}$$

*with a constant* $c_\beta>0$ *depending only on* $\beta$, $s$, $d$, $n$, $c_s$, $c_d$ $c_1$, *and* $c_2$.

*(ii) There is a continuous linear extension operator* $E_F : B^{p,p}_\beta(F)\to W^\beta_p(\mathbb{R}^n)$ *such that* $\mathrm{Tr}_F(E_F f)=f$ *for* $f\in B^{p,p}_\beta(F)$.

Theorem 6 is a particular case of [21, Theorem 1].

The spaces $B^{p,p}_\beta(F,\mu)$ are Banach spaces, while $B^{2,2}_\beta(F,\mu)$ are Hilbert spaces, and their corresponding scalar product is denoted by $\langle\cdot,\cdot\rangle_{B^{2,2}_\beta(F,\mu)}$.

A priori the definition of $B^{p,p}_\beta(F,\mu)$ depends on both $F$ and $\mu$. However, it was shown in [21, Section 3.5] that for two different measures $\mu_1$ and $\mu_2$ satisfying hypotheses of Theorem 6 and with common support $F$, if $f\in B^{p,p}_\beta(F,\mu_2)$, then $f$ can be altered on a set with $\mu_2$-measure zero, in such a way that $f$ becomes a function in $B^{p,p}_\beta(F,\mu_1)$. In other words, also by Theorem 6, the spaces $B^{2,2}_\beta(F,\mu_1)$ and $B^{p,p}_\beta(F,\mu_2)$ are equivalent. Thus, we simplify the notations and instead of $B^{p,p}_\beta(F,\mu)$ simply write $B^{p,p}_\beta(F)$.

Let us notice [21] that this time if $F$ is a $d$-set with $0<d\le n$ as defined in Sect. 3.1, then $\mu=m_d$ satisfies (4), (5), and (6), and hence, it is possible to apply Theorem 6. The restriction on $\beta$ in Theorem 6 becomes $0<\alpha<1$ with $\alpha=\beta-\frac{n-d}{p}$. Consequently, from one hand, the space $B^{p,p}_\beta(F)$ is equivalent to the Besov space $B^{p,p}_\alpha(F)$ with $0<\alpha<1$ from Sect. 3.1 (see [23]), which, from the other hand, by our previous remark for $\alpha<1$, explains why we do not need to impose that $F$ preserves the local Markov inequality. Thus in the framework of $d$-sets, this theorem coincides with Theorem 4 for $\alpha<1$.

*Remark 1* If we apply Theorem 6 for $W^{1,2}(\mathbb{R}^n)$, we obtain the image of the trace equal to the Hilbert space $B^{2,2}_1(F)$ with the restrictions

$$n\ge s\ge d>n-2\ge 0.$$

### 3.3 Integration by Parts and the Green Formula

Let us generalize the Green formula formulated for $d$-sets in [4] (initially proposed by Lancia [27, Thm. 4.15] for a von Koch curve, see also [28] for an other specific $d$-set boundary case) and the integration by parts from Appendix A Theorem A.3 [31] (see also the proof of formula (4.11) of Theorem 4.5 in [10]).

**Proposition 1 (Green Formula)** *Let $\Omega$ be a domain in $\mathbb{R}^n$ $(n \geq 2)$ with a closed boundary $\partial\Omega$ that is the support of a Borel measure $\mu$ satisfying the conditions of Theorem 6 with $n \geq s \geq d > n-2 \geq 0$. Then,*

1. *the Green formula holds for all $u$ and $v$ from $W^{1,2}(\Omega)$ with $\Delta u \in L^2(\Omega)$,*

$$\int_\Omega v\Delta u \mathrm{d}x + \int_\Omega \nabla v \cdot \nabla u \mathrm{d}x = \langle \frac{\partial u}{\partial n}, \mathrm{Tr} v\rangle_{((B_1^{2,2}(\partial\Omega))', B_1^{2,2}(\partial\Omega))}, \tag{12}$$

*where $(B_1^{2,2}(\partial\Omega))'$ is the dual space of $B_1^{2,2}(\partial\Omega)$.*

2. *In addition the usual integration by parts holds for all $u$ and $v$ from $W^{1,2}(\Omega)$ in the following weak sense:*

$$\langle u\nu_i, v\rangle_{(B_1^{2,2}(\partial\Omega))', B_1^{2,2}(\partial\Omega))} := \int_\Omega \frac{\partial u}{\partial x_i} v \mathrm{d}x + \int_\Omega u \frac{\partial v}{\partial x_i} \mathrm{d}x \quad i = 1, \ldots, n, \tag{13}$$

*where $u\nu_i$ is denoted by the linear continuous functional on $B_1^{2,2}(\partial\Omega)$.*

The statement of proposition follows, thanks to Theorem 6, from the surjectivity of the linear continuous trace operator $\mathrm{Tr}_{\partial\Omega} : W^{1,2}(\Omega) \to B_1^{2,2}(\partial\Omega)$. To prove (12) and (13), it is sufficient to follow [27] and [10], respectively. Thus the proof is omitted.

## 4 Sobolev Admissible Domains and the Generalization of the Rellich–Kondrachov Theorem

Thanks to Theorems 1 and 6, we can generalize now the notion of admissible domains introduced in [4] in the framework of $d$-sets.

**Definition 7 (Sobolev Admissible Domain)** Let $1 < p < \infty$ and $k \in \mathbb{N}^*$ be fixed. A domain $\Omega \subset \mathbb{R}^n$ is called a ($W^{k,p}$-) Sobolev admissible domain if it is an $n$-set, such that $W^{k,p}(\Omega) = C_p^k(\Omega)$ as sets with equivalent norms (hence, $\Omega$ is a $W^{k,p}$-extension domain), with a closed boundary $\partial\Omega$ that is the support of a Borel measure $\mu$ satisfying the conditions of Theorem 6.

We summarize several useful results on the trace and extension operators (see [36] for more general results for the case $p > n$) for the trace and the extension operators.

**Theorem 7** *Let $\Omega$ be a Sobolev admissible domain in $\mathbb{R}^n$, $1 < p < \infty$, $k \in \mathbb{N}^*$ be fixed and $\beta$ defined in (10). Then the following trace operators (see Definition 6)*

1. *$Tr : W^{\beta,p}(\mathbb{R}^n) \to B^{p,p}_\beta(\partial\Omega)$,*
2. *$Tr_\Omega : W^{k,p}(\mathbb{R}^n) \to W^{k,p}(\Omega)$,*
3. *$Tr_{\partial\Omega} : W^{1,p}(\Omega) \to B^{p,p}_1(\partial\Omega)$*

*are linear continuous and surjective with linear bounded right inverse, i.e. extension, operators $E : B^{p,p}_\beta(\partial\Omega) \to W^{\beta,p}(\mathbb{R}^n)$, $E_\Omega : W^{k,p}(\Omega) \to W^{k,p}(\mathbb{R}^n)$, $E_{\partial\Omega} : B^{p,p}_1(\partial\Omega) \to W^{1,p}(\Omega)$.*

***Proof*** It is a corollary of results given in Sects. 2 and 3. Indeed, if $\Omega$ is Sobolev admissible, then by Theorem 6, the trace operator $\mathrm{Tr} : W^\beta_p(\mathbb{R}^n) \to B^{p,p}_\beta(\partial\Omega) \subset L^p(\partial\Omega)$ is linear continuous and surjective with linear bounded right inverse $E : B^{p,p}_\beta(\partial\Omega) \to W^\beta_p(\mathbb{R}^n)$ (point 1). We notice that since $d \geq n-1$, the interval of suitable $\beta$, given by $]\frac{n-d}{p}, 1+\frac{n-s}{p}[$, includes as a subset $]\frac{1}{p}, 1]$ for all $p > 1$. The upper regularity limit for $\beta$ can be estimated by $1 < 1+\frac{n-s}{p} \leq 1+\frac{1}{p} < 2$.

On the other hand, by [16], $\Omega$ is a $W^{k,p}$-extension domain, and $\mathrm{Tr}_\Omega : W^{k,p}(\mathbb{R}^n) \to W^{k,p}(\Omega)$ and $E_\Omega : W^{k,p}(\Omega) \to W^{k,p}(\mathbb{R}^n)$ are linear continuous (point 2). Hence, the embeddings for $k = 1$

$$B^{p,p}_1(\partial\Omega) \to W^{1,p}(\mathbb{R}^n) \to W^{1,p}(\Omega) \quad \text{and} \quad W^{1,p}(\Omega) \to W^{1,p}(\mathbb{R}^n) \to B^{p,p}_1(\partial\Omega)$$

are linear continuous (point 3). □

By updating the class of admissible domains, all results of [4] still hold in this new class. For instance, it is also possible to consider Sobolev admissible truncated domains for which two disjoint boundaries satisfy Theorem 6. Without any particular motivation here for the truncated domain, let us just formulate the compactness of the embedding $W^{1,2}(\Omega)$ to $L^2(\Omega)$ for the Sobolev admissible domains.

**Proposition 2** *Let $\Omega$ be a bounded Sobolev admissible domain for $p = 2$ and $k = 1$. Then the Sobolev space $W^{1,2}(\Omega)$ is compactly embedded in $L^2(\Omega)$:*

$$W^{1,2}(\Omega) \subset\subset L^2(\Omega).$$

The proof follows with small modifications the proof of Proposition 2 in [4] and thus is omitted.

However, we would like to recall main compactness results of [4] putting them in the new framework of Sobolev admissible domains with not necessarily a $d$-set boundary.

*Remark 2* To have a compact embedding, it is important that the domain $\Omega$ be a bounded $W_p^k$-extension domain.

As a direct corollary, we have the following generalization of the classical Rellich–Kondrachov theorem (see, for instance, Adams [2] p.144 Theorem 6.2).

**Theorem 8 (Compact Sobolev Embeddings for $n$-Sets, [4])** *Let $\Omega \subset \mathbb{R}^n$ be a bounded n-set with $W^{k,p}(\Omega) = C_p^k(\Omega)$, $1 < p < \infty$, $k, \ell \in \mathbb{N}^*$. Then the following embeddings are compact:*

1. $W_p^{k+\ell}(\Omega) \subset\subset W_q^\ell(\Omega)$,
2. $W_p^k(\Omega) \subset\subset L^q(\Omega)$,

*with $q \in [1, +\infty[$ if $kp = n$, $q \in [1, +\infty]$ if $kp > n$, and with $q \in [1, \frac{pn}{n-kp}[$ if $kp < n$.*

## 5 Compactness of the Trace

We generalize the classical Rellich–Kondrachov theorem for fractals in the following way.

**Theorem 9 (Compact Besov Embeddings)** *Let $F \subset \mathbb{R}^n$ be a compact set satisfying conditions of Theorem 6 with $\beta = 1$. Assume $1 \leq q \leq p < \infty$.*

*Then,*

1. *the linear trace operator* Tr : $W^{1,p}(\mathbb{R}^n) \to L^q(F)$ *is compact,*
2. *the continuous embedding $B_1^{p,p}(F) \subset\subset L^q(F)$ is compact.*

***Proof*** Indeed, for $\beta = 1$, thanks to Theorem 6, the extension $E_F : B_1^{p,p}(F) \to W_p^1(\mathbb{R}^n)$ is continuous. Thus, if the trace operator Tr : $W^{1,p}(\mathbb{R}^n) \to L^q(F)$ is compact, by the composition of the bounded and compact operators, the embedding $B_1^{p,p}(F) \subset\subset L^q(F)$ is compact too.

Hence, let us prove the first statement of the theorem. Taking, for instance, a non-trivial ball $B_r \subset \mathbb{R}^n$, or, more generally, any extension domain, in the way that $F \subset \overline{B_r}$, we notice that, as the operator $W^{1,p}(\mathbb{R}^n) \to W^{1,p}(B_r)$ is continuous, the question can be reduced to the compactness of $\mathrm{Tr}_{B_r,F} : W^{1,p}(B_r) \to L^q(F)$.

Using the idea of the proof of [3, Proposition 8.1], we take a bounded sequence $(u_n)_{n\in\mathbb{N}}$ of $W^{1,p}(B_r)$. As $B_r$ is a bounded extension domain, then by point 2 of Theorem 8 (here, as $\partial B_r$ is $C^\infty$, even by the classical Rellich–Kondrachov theorem), the inclusion $W^{1,p}(B_r) \to L^p(B_r)$ is compact. Taking $1 \leq q \leq p$, we have the usual continuous embedding $L^p(F) \subset L^q(F)$, holding actually for all bounded sets.

Therefore, there exists a subsequence $(u_{n_k})_{k\in\mathbb{N}} \subset (u_n)_{n\in\mathbb{N}}$ with an element $u \in W^{1,p}(B_r)$ such that

$$u_{n_k} \rightharpoonup u \text{ in } W^{1,p}(B_r), \quad u_{n_k} \to u \text{ in } L^p(B_r) \text{ for } k \to +\infty.$$

Moreover, still by the boundness of $(u_{n_k})_{k\in\mathbb{N}}$ in $W^{1,p}(B_r)$, there exists a constant $M > 0$ such that

$$\forall k \in \mathbb{N} \quad \|\nabla u_{n_k}\|^p_{L^p(B_r)} \leq M.$$

Since $B_1^{p,p}(F) \subset L^p(F) \subset L^q(F)$, combining the previously viewed continuous embeddings and Theorem 6, we obtain that the linear operator $\mathrm{Tr}_{B_r,F} : W^{1,p}(B_r) \to L^q(F)$ is bounded.

Consequently, the sequence $(\mathrm{Tr}_{B_r,F} u_{n_k})_{k\in\mathbb{N}}$ is bounded in $L^q(F)$ with $1 \leq q \leq p$, and thus in addition (passing if necessarily again to a subsequence, denoted again by $(\mathrm{Tr}_{B_r,F} u_{n_k})_{k\in\mathbb{N}}$) $\mathrm{Tr}_{B_r,F} u_{n_k} \rightharpoonup w$ in $L^q(F)$ for an element $w \in L^q(F)$ (we will see in the following that $w = \mathrm{Tr}_{B_r,F} u$). Without loss of generality, let us assume that $u = 0$ and, hence, let us prove that $\mathrm{Tr}_{B_r,F} u_{n_k} \to 0$ in $L^q(F)$.

We firstly notice that for all $\sigma > 0$ there exists a constant $c = c(\sigma) > 0$ such that for all $u \in W^{1,p}(\Omega)$ it holds

$$\|\mathrm{Tr}_{B_r,F} u\|^p_{L^q(F)} \leq \frac{1}{\sigma}\|\nabla u\|^p_{L^p(B_r)} + c(\sigma)\|u\|^p_{L^p(B_r)}. \tag{14}$$

Actually, by the boudness of $(u_{n_k})_{k\in\mathbb{N}}$ in $W^{1,p}(\Omega)$, for all $\delta \in [0, 1]$ there exists a constant $c(\delta, \hat{M}) > 0$ such that for all $k \in \mathbb{N}$

$$\left\|\nabla u_{n_k}\right\|^p_{L^p(B_r)} + \left\|u_{n_k}\right\|^p_{L^p(B_r)} \leq \delta \left\|\nabla u_{n_k}\right\|^p_{L^p(B_r)} + c(\delta, \hat{M}) \left\|u_{n_k}\right\|^p_{L^p(B_r)}.$$

By the continuity of $\mathrm{Tr}_{B_r,F} : W^{1,p}(B_r) \to L^q(F)$, there exists a constant $C_{tr} > 0$ (depending on all constants from Theorem 6) such that for all $k \in \mathbb{N}$

$$\left\|\mathrm{Tr}_{B_r,F} u_{n_k}\right\|^p_{L^q(F)} \leq C_{tr}\delta \left\|\nabla u_{n_k}\right\|^p_{L^p(B_r)} + C_{tr} c(\delta, \hat{M}) \left\|u_{n_k}\right\|^p_{L^p(B_r)}.$$

Thus we denote $\sigma = C_{tr}\delta > 0$ which can be arbitrary small thanks to the arbitrariness on $[0, 1]$ of $\delta$ and obtain (14). Let us define now $\varepsilon = \frac{M}{\sigma} > 0$ for $M > 0$, the boundness constant of $\nabla u_{n_k}$ in $L^p(B_r)$. Then for all $k \in \mathbb{N}$, it holds

$$\|\mathrm{Tr}_{B_r,F} u_{n_k}\|^p_{L^q(F)} \leq \varepsilon + c\|u_{n_k}\|^p_{L^p(B_r)}$$

with a constant $c > 0$ independent on $k$. Since $\|u_{n_k}\|_{L^p(B_r)} \to 0$ for $k \to +\infty$, we find that for all $\varepsilon > 0$ $\limsup_{k\to+\infty} \|\mathrm{Tr}_{B_r,F} u_{n_k}\|^p_{L^q(F)} \leq \varepsilon$, which directly ensures that $\mathrm{Tr}_{B_r,F} u_{n_k} \to 0$ in $L^q(F)$. This finishes the proof of point 2. □

*Remark 3* The statement of point 2 [4, Theorem 2.13] for compact $d$-set preserving Markov's local inequality should be updated by the general result on the compactness of Besov embeddings on $d$-sets from [38, Proposition 20.5, Theorem 20.6], which actually contains the case $q \in [1, p]$ of Theorem 9 independently on the value of the dimension on the value of the dimension $d$.

A complement result about compacteness of the embeddings $W^{1,p}(\Omega) \to L^q(F)$ for an extension domain $\Omega$ and $F \subset \Omega$, defined as a support of a measure satisfiyng only the upper estimate in the d-set condition with $0 \leq n - p \leq d \leq n$, is given in [6]: for $p = n \geq 2$ with $q \in [1, +\infty]$, for $p > n$ and $p = n = 1$ with $q = \infty$, and finaly, for $1 < p < n$ and $s = dp/n - p > p$ for $q \in [1, s]$. We also notice that working in the framework of Theorem 6, if $p = +\infty$, then $\beta$ is necessarily strictly less than 1.

In particular, the compactness of the trace operator implies the following equivalence of the norms on $W^{1,p}(\Omega)$.

**Proposition 3** *Let $\Omega$ be a Sobolev admissible domain in $\mathbb{R}^n$ with a compact boundary $\partial\Omega$ and $1 < p < \infty$, $k \in \mathbb{N}^*$. Then,*

1. $W^{k,p}(\Omega) \subset\subset L^p_{loc}(\Omega)$, $k \in \mathbb{N}^*$;
2. $\mathrm{Tr} : W^{1,p}(\Omega) \to L^p(\partial\Omega)$ *is compact;*
3. *if in addition the measure $\mu$ is Borel regular, then the image* $\mathrm{Im}(\mathrm{Tr}) = B^{p,p}_1(\partial\Omega)$ *is dense in $L^p(\partial\Omega)$;*
4. $\|u\|_{W^{1,p}(\Omega)}$ *is equivalent to* $\|u\|_{\mathrm{Tr}} = \left(\int_\Omega |\nabla u|^p dx + \int_{\partial\Omega} |\mathrm{Tr} u|^p d\mu\right)^{\frac{1}{p}}$.

***Proof*** Let us prove point 3. To prove all other points, it is sufficient to follow the proof of Proposition 3 in [4].

If $\partial\Omega$ is endowed with a Borel regular measure $\mu$, then the space $\{v|_{\partial\Omega} : v \in \mathcal{D}(\mathbb{R}^n)\}$, which is dense in $C(\partial\Omega)$ by the Stone–Weierstrass theorem for the uniform norm, is also dense in $L^p(\partial\Omega)$ (see Theorem 2.11 in [13]). Hence, $B^{p,p}_k(\partial\Omega)$ is dense in $L^p(\partial\Omega)$. □

The Poincaré's inequality stays also true on a bounded Sobolev admissible domain [11].

**Theorem 10 (Poincaré's Inequality)** *Let $\Omega \subset \mathbb{R}^n$ with $n \geq 2$ be a bounded connected Sobolev admissible domain. For all $u \in W^{1,p}_0(\Omega)$ with $1 \leq p < +\infty$, there exists $C > 0$ depending only on $\Omega$, $p$, and $n$ such that*

$$\|u\|_{L^p(\Omega)} \leq C\|\nabla u\|_{L^p(\Omega)}.$$

*Therefore the seminorm $\|.\|_{W^{1,p}_0(\Omega)}$, defined by $\|u\|_{W^{1,p}_0(\Omega)} := \|\nabla u\|_{L^p(\Omega)}$, is a norm that is equivalent to $\|.\|_{W^{1,p}(\Omega)}$ on $W^{1,p}_0(\Omega)$.*

*Moreover for all $u \in W^{1,p}(\Omega)$, there exists $C > 0$ depending only on $\Omega$, $p$, and $n$ such that*

$$\left\| u - \frac{1}{\lambda(\Omega)} \int_\Omega u \, d\lambda \right\|_{L^p(\Omega)} \leq C \|\nabla u\|_{L^p(\Omega)}.$$

***Proof*** The result for $u \in W_0^{1,p}(\Omega)$ comes from the boundness of $\Omega$. The result for $u \in W^{1,p}(\Omega)$ comes from the compactness of the embedding $W^{1,p}(\Omega) \subset\subset L^p(\Omega)$ from Theorem 8 and following for instance the proof from [13] (see section 5.8.1 Theorem 1). □

Thus the results of [4] on the Dirichlet-to-Neumann operator can also be updated in the framework of the Sobolev admissible domains of Definition 7. For instance, we have:

**Theorem 11** *Let $\Omega$ be a bounded Sobolev admissible domain in $\mathbb{R}^n$ $(n \geq 2)$ for $p = 2$ and $k = 1$. Then the Poincaré-Steklov operator*

$$A : B_1^{2,2}(\partial\Omega) \to (B_1^{2,2}(\partial\Omega))'$$

*mapping $u|_{\partial\Omega}$ to $\partial_\nu u|_{\partial\Omega}$ is a linear bounded self-adjoint operator with* $\ker A \neq 0$.

# 6 Application to the Poisson Boundary Valued and Spectral Problems

In this section we show the application of the theory of functional spaces developed in the previous sections on the example of the Poisson equation with Robin boundary conditions that we can weakly solve on the Sobolev admissible domains.

Let $\Omega$ be a $W^{1,2}$-Sobolev admissible domain with a compact boundary $\partial\Omega$ and $f \in L^2(\Omega)$. For $a > 0$, we define $W^{1,2}(\Omega)$ endowed with the equivalent by Proposition 3 norm

$$\|u\|_{\mathrm{Tr}}^2 = \int_\Omega |\nabla u|^2 dx + a \int_{\partial\Omega} |Tr_{\partial\Omega} u|^2 d\mu. \tag{15}$$

Then $u \in W^{1,2}(\Omega)$ is called a weak solution of the Poisson problem (1) if for all $v \in W^{1,2}(\Omega)$

$$(u, v)_{\mathrm{Tr}} = \int_\Omega \nabla u \nabla v \, dx + a \int_{\partial\Omega} Tr_{\partial\Omega} u Tr_{\partial\Omega} v \, d\mu = \int_\Omega f v \, dx.$$

Thus, the Riesz representation theorem gives us the well-posedness result.

**Theorem 12** *Let $\Omega$ be a $W^{1,2}$-Sobolev admissible domain with a compact boundary $\partial\Omega$. Then for all $f \in L^2(\Omega)$ and $a > 0$, there exists a unique weak solution $u \in W^{1,2}(\Omega)$ of the Poisson problem (1), and it holds the stability estimate*

$$\|u\|_{\mathrm{Tr}} \leq C\|f\|_{L^2(\Omega)}.$$

In the same time with the additional assumption that $\Omega$ is bounded, ensuring the compactness of the embedding $i_{L^2(\Omega)} : W^{1,2}(\Omega) \to L^2(\Omega)$ by Proposition 3, we also have the compactness of the operator $B : f \in L^2(\Omega) \mapsto B(f) = u \in W^{1,2}(\Omega)$ mapping a source term $f$ to the weak solution of the Poisson problem (1) (see, for instance, Theorem 3.6 [4]). The compactness of the embedding $i_{L^2(\Omega)}$ also allows to apply the spectral Hilbert–Schmidt theorem for a self-adjoint compact operator on a Hilbert space to obtain the usual properties of the spectral problem for the $-\Delta$ on the Sobolev admissible domains.

**Theorem 13** *Let $\Omega$ be a bounded $W^{1,2}$-Sobolev admissible domain. The weak eigenvalue problem*

$$\forall v \in W^{1,2}(\Omega) \quad (u, v)_{\mathrm{Tr}} = \lambda(u, v)_{L^2(\Omega)}$$

*has a countable number of strictly positive eigenvalues of finite multiplicity, which is possible to numerate in the nondecreasing way*

$$0 < \lambda_1 \leq \lambda_2 \leq \lambda_3 \leq \cdots, \quad \lambda_j \to +\infty \quad j \to +\infty.$$

*In addition the corresponding eigenfunctions form an orthonormal basis of $L^2(\Omega)$ and an orthogonal basis of $W^{1,2}(\Omega)$.*

***Proof*** It is sufficient to notice that the eigenvalue problem is equivalent to the spectrum problem $Tu = \frac{1}{\lambda}u$ for the operator $T = A \circ i_{L^2(\Omega)} : W^{1,2}(\Omega) \to W^{1,2}(\Omega)$, which is linear compact and self-adjoint on the Hilbert space $W^{1,2}(\Omega)$. Here $A$ is the linear bounded operator (existing by the Riesz representation theorem) that maps $v \in L^2(\Omega)$ to $Av \in W^{1,2}(\Omega)$ such that

$$\forall \phi \in W^{1,2}(\Omega) \quad (v, \phi)_{L^2(\Omega)} = (Av, \phi)_{\mathrm{Tr}}.$$

□

## References

1. Achdou, Y., Deheuvels, T., Tchou, N.: Comparison of different definitions of traces for a class of ramified domains with self-similar fractal boundaries. Potential Anal. **40**, 345–362 (2013)
2. Adams, R.A., Fournier, J.J.F.: Sobolev Spaces. Academic Press, Cambridge (2003)

3. Arendt, W., ter Elst, A.: The Dirichlet-to-Neumann operator on rough domains. J. Differ. Equ. **251**, 2100–2124 (2011)
4. Arfi, K., Rozanova-Pierrat, A.: Dirichlet-to-Neumann or Poincaré-Steklov operator on fractals described by d-sets. Discrete Contin. Dynam. Systems **12**, 1–26 (2019)
5. Bardos, C., Grebenkov, D., Rozanova-Pierrat, A.: Short-time heat diffusion in compact domains with discontinuous transmission boundary conditions. Math. Models Methods Appl. Sci. **26**, 59–110 (2016)
6. Biegert, M.: On traces of Sobolev functions on the boundary of extension domains, Proc. Am. Math. Soc. **137**, 4169–4176 (2009)
7. Bos, L.P., Milman, P.D.L Sobolev-Gagliardo-Nirenberg and Markov type inequalities on subanalytic domains. Geom. Funct. Anal. **5**, 853–923 (1995)
8. Calderon, A.-P.: Lebesgue spaces of differentiable functions and distributions. Proc. Symp. Pure Math. **4**, 33–49 (1961)
9. Capitanelli, R.: Asymptotics for mixed Dirichlet–Robin problems in irregular domains. J. Math. Anal. Appl. **362**, 450–459 (2010)
10. Creo, S., Lancia, M.R., Vernole, P., Hinz, M., Teplyaev, A.: Magnetostatic problems in fractal domains (2018). arXiv:1805.08262
11. Dekkers, A.: Mathematical analysis of the Kuznetsov equation: Cauchy problem, approximation questions and problems with fractals boundaries. Ph.D. Thesis, CentraleSupélec, Université Paris Saclay (2019)
12. Edmunds, D., Evans, W.: Spectral theory and differential operators, Oxford Mathematical Monographs. Oxford University Press, Oxford (1987)
13. Evans, L.C.: Partial Differential Equations. American Math Society, Providence (2010)
14. Girault, V., Raviart, P.-A.: Finite Element Methods for the Navier-Stokes Equations, Theory and Algorithms. Springer, New York (1986)
15. Grisvard, P.: Théorèmes de traces relatifs à un polyèdre. C.R. Acad. Sci. Paris Sér. A **278**, 1581–1583 (1974)
16. Hajlasz, P., Koskela, P., Tuominen, H.: Sobolev embeddings, extensions and measure density condition. J. Funct. Anal. **254**, 1217–1234 (2008)
17. Herron, D.A., Koskela, P.: Uniform, Sobolev extension and quasiconformal circle domains. J. Anal. Math. **57**, , 172–202 (1991)
18. Hinz, M., Rozanova-Pierrat, A., Teplyaev, A.: Fractal shape optimization with applications to linear acoustics. SIAM J. Control Optim. to appear, arXiv:2008.10222 (2020)
19. Ihnatsyeva, L., Vähäkangas, A.V.: Characterization of traces of smooth functions on Ahlfors regular sets (2011). arXiv:1109.2248
20. Jones, P.W.: Quasi conformal mappings and extendability of functions in Sobolev spaces. Acta Math. **147**, 71–88 (1981)
21. Jonsson, A.: Besov spaces on closed subsets of $\mathbb{R}^n$. Trans. Am. Math. Soc. **341**, 355–370 (1994)
22. Jonsson, A.: Besov spaces on closed sets by means of atomic decomposition. Comp. Variab. Elliptic Equ. **54**, 585–611 (2009)
23. Jonsson, A., Wallin, H.: Function spaces on subsets of $\mathbb{R}^n$. Math. Reports 2, Part 1, Harwood Academic Publishers, London (1984)
24. Jonsson, A., Wallin, H.: The dual of Besov spaces on fractals. Studia Math. **112**, 285–300 (1995)
25. Jonsson, A., Wallin, H.: Boundary value problems and Brownian motion on fractals. Chaos, Solitons Fractals **8**, 191–205 (1997)
26. Jonsson, A., Sjögren, P., Wallin, H.: Hardy and Lipschitz spaces on subsets of $\mathbb{R}^n$. Studia Math. **80**, 141–166 (1984)
27. Lancia, M.R.: A transmission problem with a fractal interface. Zeitschrift für Analysis und ihre Anwendungen **21**, 113–133 (2002)
28. Lancia, M.R.: Second order transmission problems across a fractal surface, Rendiconti, Accademia Nazionale delle Scienze detta dei XL. Memoire di Mathematica e Applicazioni **XXVII**, 191–213 (2003)

29. Lancia, M.R., Vernole, P.: Irregular heat flow problems. SIAM J. Math. Anal. **42**, 1539–1567 (2010)
30. Lions, J., Magenes, E.: Non-Homogeneous Boundary Value Problems and Applications, vol. 1. Springer, Berlin (1972)
31. Magoulès, F., Nguyen, T.P.K., Omnès, P., Rozanova-Pierrat, A.: Optimal absorption of acoustical waves by a boundary. Part 2 Submitted (2017)
32. Marschall, J.: The trace of Sobolev-Slobodeckij spaces on Lipschitz domains. Manuscripta Math **58**, 47–65 (1987)
33. Martio, O., Sarvas, J.: Injectivity theorems in plane and space. Annal. Acad. Scientiarum Fennicae Series A I Mathematica **4**, 383–401 (1979)
34. Necas, J.: Les Méthodes Directes en Théorie des Équations Elliptiques. Masson, Paris (1967)
35. Raviart, P.-A., Thomas, J.-M.: Introduction à l'analyse numérique des équations aux dérivées partielles. Masson, Paris (1983)
36. Shvartsman, P.: On the boundary values of Sobolev $W_p^1$-functions. Adv. Math. **225**, 2162–2221 (2010)
37. Stein, E.M.: Singular Integrals and Differentiability Properties of Functions. Princeton University Press, Princeton (1970)
38. Triebel, H.: Fractals and Spectra. Related to Fourier Analysis and Function Spaces. Birkhäuser, Basel (1997)
39. Wallin, H.: Markov's inequality on subsets of $\mathbb{R}^n$, 6. Department of Mathematics, University of Umea (1982)
40. Wallin, H.: The trace to the boundary of Sobolev spaces on a snowflake. Manuscripta Math. **73**, 117–125 (1991)
41. Wingren, P.: Lipschitz Spaces and Interpolating Polynomials on Subsets of Euclidean Space, pp. 424–435. Springer Science + Business Media, Berlin (1988)

Printed in the United States
by Baker & Taylor Publisher Services